Lovebug

Daisy Lafarge

P
PENINSULA PRESS, LONDON

For those who lose their heads

Contents

⟲	Let's Hook Up	11
⇢	Love Kills People	31
⌘	The Bite	49
✪	I Lose My Head	75
⨌	We Eat Each Other Up	101
⟳	A Species Is An Idea	125

Acknowledgements   149
Images   151
References   153

I call it *pathogen* because it promises to make me suffer. Love turns my friends into doctors and some of them say *vigilance* and others say *supplements*. I take all the borders I am prescribed. The pathogen arrives anyway and takes a seat at the table. Conditioned to welcome damage, I am curious about this uninvited guest. *You must sit down, says Love, and taste my meat.* We become acquainted – for a while things are thrilling, I forget myself, I forget to eat, I send nudes, barely sleep. My sweat sharpens, and then shifts to something metallic. The pathogen is changing too – behaving differently, erratically, misplacing their parts, strands of DNA mixed in with mine. I fuck with theirs too, it's pathogen play; we take turns being the infection and our cells are so into it. My tastes in pathology are changing. I stop cruising for damage in an expectation of harmony. I stop using pathogen like *essence* and start meaning it like *variability* or *something I'm into right now* or *risk from which I might not emerge intact*, like any kind of love. I start seeing pathogen like a mirror of being alive. *So I did sit and eat.*

Let's Hook Up

First, single-celled animals signaled to each other, "I'm going to devour you," "I'm over here!" "Let's hook up," etc.

– Eleni Sikelianos

This book is an exercise in abjection. By the time I became aware of this it was already too late. There were clues from the beginning, which I ignored: people's reactions to what I was spending my time on, my obsessive collecting of terminology and anecdotes, the conceptual contortions I was unable to perform but considered necessary for the task. The task being to write about the seemingly unwritable: bacteria, viruses, parasites and infections, specifically those which cross species boundaries between animals and humans. The pandemic was still four years in the future so there was no obvious timeliness at play; neither was I a scientist or a journalist or, at that point, significantly ill. Behind people's immediate disgust or bafflement at my choice of subject I sometimes sensed a judgement of my own morbidity: what pulled me to write about infectious life, if not a kind of perversion?

Poets are known – even expected – to write about nature and its nominal updates: the environment, ecology, the nonhuman, the Anthropocene. Not that this has ever been straightforward. Nature can be the poet's redeemer or lover or companion or punisher. For a long time it was a kind of frenemy. Then came the twenty-first century: Baudelaire was dead but all the *fleurs* were *mal*. Microplastics in the air and fatbergs in the sewer. It had always been possible to contract a virus from a kiss, but now we knew exactly how it worked. This was all nature and I couldn't look away. Couldn't, but also – didn't want to, since I was still at the mercy of a lapsed evangelical attraction to conceptual underdogs, of a desire to gaze with fearful love on Job's infected wounds. Trees are sublime, but have you ever watched the clumsy arabesques of a paramecium as it glides through a drop of pondwater?

In the middle of my twenties I saw a description for a research scholarship, advertised by a university close to where I lived. They wanted a writer who was interested in zoonoses. After googling the unfamiliar word I knew I would apply. I wanted to write about nature but not the uncomplicated kind; I was drawn to where things got ugly, murky, difficult, potentially threatening. This was less a conscious masochism and more like an autoimmune response that mistook proximity for safety. Over the next few years I shadowed epidemiologists in Glasgow and Edinburgh, and spent time in a zoonoses lab at the foot of Mount Kilimanjaro. I attended conferences, meetings, lab demonstrations, animal vaccinations and dissections.

I spoke to vets, farmers, scientists, anthropologists, statisticians, friends, strangers, other writers and artists. I tried to be good; I studied microbiology. I read papers and submerged myself in their terminology, in which microorganisms and their discontents were rendered in dense Latin. I tried to be good, but I never achieved the expertise I thought would finally validate me to write about infectious life, the uneasy bonds that tie all living things together.

When I describe this book as an exercise in abjection it is not a phrase I use lightly. In Julia Kristeva's *Powers of Horror* abjection is the reality-altering horror and revulsion experienced when an individual's sense of self is presented with something that threatens its coherence. Often this something is a state of matter that confronts the self with its own materiality: shit, piss, vomit, pus, open wounds, orifices, worms, maggots, beetles, anything in a state of decay or decomposition, like the rotting flesh of cadavers, or visible infection and disease. Abjection cracks open the self, muddying the distinction between human and nonhuman, life and death. In response, the self must employ a principle of exclusion – it must reject and cast out threats in order to regain an illusory coherence. Cue sanitation and hygiene, but also, according to Kristeva, phobias and compulsive behaviours.

The subject of my research, then, would seem a neat fit. Infections threaten my individual coherence not only by threatening my life or wellbeing, but also by showing blatant disregard for the illusion of my own integrity.

Infections do not ask to become a part of me before they do it. They come through a bite or a scratch or a cough or a kiss, through air or through skin. Often I am the last to know about it, and by the time I do I am already – as the clinical language would have it – colonised. It made sense to me when my friend Kirsty, an artist, told me that 'bug' (likely from the Middle English *bugge*) was originally a word for something frightening, long before it became a term for an insect. Later, its specific use in relation to insects replaced its general application for something threatening. This older bug or *bugge* is maybe a better word than abjection, since it conflates the feeling provoked (horror, fear) with the thing that caused it. Bundled up together, cause and effect are subject to the loss of coherence and linearity that typifies abjection.

But that's not the only reason I've come to think of this book in abject terms. *Powers of Horror* connects abjection with language, since language is often enlisted in an attempt to shore up the border laid bare by the experience of abjection. An obsessive relation to words – possibly detached from their meanings – consumes a frightened self that cocoons in a surrogate structure of letters, syllables and symbols. If I find myself recoiling at a surface reading of this text – which would seem to pathologize an attachment to language – I also find it relatable. Faced with indeterminacy, I am often reduced to anxious inertia. A blank page, a complex subject, a situation with no clear right or wrong. I understand this as the result of being raised with a deficit of critical faculties: in the evangelical Baptist churches where I spent large periods of my childhood, there was no need to cultivate a capacity for ambiguity and contradiction. There was good and then there was evil – the world was a neat binary of right and wrong.

Emerging from this context into a kind of epistemically impaired adulthood, I had only a single tool at my disposal. I collected words and anecdotes as if my life depended on it. I could memorise textbooks and regurgitate them word for word in the following day's exam, retaining nothing a week later. Terrified by the edgelessness of a new project, I would quickly fill pages and pages with research. There was pleasure mixed in with this panic, too – a woozy logophilia that knitted me into the work, a not unerotic thrill when resonance and serendipity began to appear in the material I had collected. In this way, the flawed mysticism and glossolalia of my upbringing lived on, finding connections and covert languages where often, 'objectively', there were none.

My parasitic methods of research – a failure to comply with the boundaries of knowledge and its appeal to objectivity – seemed to echo the microorganisms I was studying. Pathogens do not treat the body as a closed text: they are a painful reminder of our openness to the world and each other, leading philosopher Donna Haraway to describe life as 'a window of vulnerability'. In view of windows, the word *knowledge* comes apart. There is the structure of what we know and then there are its ledges, an offing into uncertainty.

If abjection can be loosely described as the state of finding parts of oneself intolerable, or experiencing a lack of borders between self and environment, then its working through requires learning to love what is difficult, learning to repair the murderous parts, as theorist Eve Kosofsky

Segdwick puts it, 'into something like a whole'. Abjection's porosity between inside and outside, self and nonself, is the most useful way I have found of thinking about pathogenic life that can become, to varying extents, a part of me. How can I learn to accept the aspects of myself – or what becomes part of myself – that may also cause me harm? To repair the murderous parts into something like a whole?

The more time I spent immersed in discourses of infection, the more I found that the dominant ways of thinking and speaking were still reading from an old script about good versus evil – an evil that intrudes on and infects the good; the pure self of an immune system, threatened by pathogenic others. This so-called military metaphor is still, to a large extent, the status quo. Even in the post-Pasteurian present of probiotics, fermented foods, faecal transplants and flourishing microbiomes, the newfound acceptance of some bugs as 'good' depends on the reinscribing of others as 'bad'. It came as no surprise when, in the early months of 2020, Boris Johnson described the novel coronavirus as an 'invisible mugger' that had to be 'wrestled to the floor', as if the virus was a delinquent teenager in breach of its ASBO.

I want to get the military metaphor out of the way because it bores me. The war on germs, the body's defences, the immune system's resistance to pathogenic intruders. Infection = invasion, a neat metaphor that maintains a divide between health and sickness, good and bad, self and other. Susan Sontag went famously hard on it, and on metaphorical thinking in relation to illness in general. *Illness as Metaphor* was written in response to a diagnosis of breast cancer, when Sontag found herself embroiled in and frustrated by the mystifications surrounding the illness. Cancer, she wrote, was too frequently used as a metaphor for the ills of society, whereas perceptions of the illness itself were clouded by fatalism and misinformation, such as the view that cancer patients could, with the right affirmations, think their way out of it. Since metaphors and myths around illness were not only unhelpful, but impacted how and if people sought treatment, Sontag was convinced that metaphors could kill.

The military metaphor's origins are difficult to trace, but the widespread acceptance of germ theory is one place to start, since it replaced traditional equations of illness with sin – disease as divine punishment or prompt. Glimmers of germ theory had been emerging since Roman times, but it wasn't proven until the late nineteenth century, when the chemist-turned-microbiologist Louis Pasteur demonstrated that certain bacteria were responsible for transmitting infectious diseases. Pasteur's discoveries were something of a poisoned chalice: on the one hand we are indebted to them, and his subsequent work on vaccines. On the other, this inheritance came tied up with Pasteur's phobic attitude towards microbial life. His biographer includes an anecdote by Monsieur Loir, the Pasteur family's regular dinner guest:

> At the dinner table [Pasteur] would wipe glassware and dinnerware in the hope of removing contaminating dirt. [Monsieur] Loir has described the odd behavior

that arose from his habit of intense and detailed observation. 'He minutely inspected the bread that was served to him and placed on the tablecloth everything he found in it: small fragments of wool, of roaches, of flour worms. Often I tried to find in my own piece of bread from the same loaf the objects found by Pasteur, but could not discover anything. All the others ate the same bread without finding anything in it. This search took place at almost every meal and is perhaps the most extraordinary memory that I have kept of Pasteur.

Medical professionals were suspicious of Pasteur's germ phobias, and his contemporary Florence Nightingale did not believe in the concept of infection, dismissing it as a 'germ-fetish'. And for Pasteur it was a fetish, albeit one that went hand in hand with another fetishistic aversion: his horror of democracy and hatred of the working classes. Born in the wake of the Reign of Terror, whatever the aristocratic Pasteur saw when he looked down the microscope was mediated by his own right-wing politics. Down there was anarchy, a swarming bacterial mob endlessly reproducing and repopulating itself. Just as the urban mobs threatened to overthrow French sovereignty and damage its integrity, the microbial mob was primed to destroy the human body at any given moment. And if the language to describe both problematic populations was the same, so too was that of the cure: total mastery and immobilisation.

By the time Sontag's book arrived, this perception of microbial life had become naturalised into a 'war' against germs and disease. In his essay 'On Truth and Lies in a Nonmoral Sense', Friedrich Nietzsche wrote about such 'worn-out metaphors which have become powerless to affect the senses,' calling them 'coins which have their obverse effaced and now are no longer of account as coins but merely as metal.' Sontag's capacity for polemicising has a consistently tiring effect on me, but never more so than in *Illness as Metaphor*, which rails against metaphor with a belligerence I find surprising from someone who has made a life in language. I first read it on my way to a conference about zoonotic diseases, where I would join scientists, policymakers, vets and statisticians. It was my second or third time attending such conferences, each one instilling a feeling that never quite ebbed; not only of being out of place, as a writer among medical professionals, but of representing something fundamentally shallow and frivolous, and, in its self-indulgence, offensive.

I wish I had read novelist Cynthia Ozick's essay 'Metaphor and Memory' before attending these conferences. Ozick's account of being called to speak in front of an assembly of doctors painfully struck home:

> Yet the writer is cautious, even frightened. Here among the doctors, the redemptive ardor of literature begins to take on a vanity. How frivolous it seems, how trivial – vanity of vanities! The doctors are absorbed by blood and bone; each one, alone in his judgment, walks the fragile bridge between the salvation into life and the morbid slide toward death. The writer is as innocent as a privileged child before all this, a sybarite of libraries, a voluptuary of print [...] What gall, to suppose that a dreamer of tales can bring news of the human predicament to the doctors on their dread rounds!

Though I felt unable to articulate it at the time, I knew that the 'baubles' of figurative language that Ozick describes are not just frippery – they are inseparable from experience, memory and history, from what makes it possible, in Ozick's words, 'to envision the stranger's heart'. Much too late I realised my feelings of shame at these conferences had more to do with resentment. Writers who find themselves in interdisciplinary contexts will perhaps recognise the predicament: to be paralysed and dismayed by your uselessness, and simultaneously resentful of the expectation that you should have to be useful at all. My presence at the conference, and my project as a whole, had been funded on this assumption of writing's utility.

These days I am less squeamish about agreeing with Bataille ('Literature is not innocent. It is guilty and should admit itself so') and Baudelaire ('To be a useful man has always seemed to me ghastly'). Ozick's *envisioning the stranger's heart* now sounds – to me – too close to virtue, or what might be rehashed in a funding application as *useful* or *socially engaged*. Literature *is* evil, if it's doing anything interesting, in the sense that it's usually unprofitable and unproductive, a waste of money and time, two things that capitalism cannot abide. Literature is necessary, not because it is useful, but because it is luxurious. It holds space for a world untethered from alienated productivity, in which killing time is a universal luxury.

If literature is evil, then metaphor would appear to be its destroying angel. Metaphor is the zenith of indulgence, excess, lushness and ornament. As much as I agree with Sontag's critique of the military metaphor, I find her overall argument hard to disentangle from a classical disdain for figurative language. In his first-century work of literary criticism *On the Sublime*, Longinus warns that metaphors are liable to lead to 'excess'; he notes that Caecilius suggests two or three at most per text, while Aristotle and Theophrastus advise that metaphors be softened with phrases like *as if* and *as it were*. In these examples, metaphors are permitted so long as they are kept on a tight leash. A similar chokehold persists in the praising of non-figurative prose as clean, refined, taut, clear, economical and clinical, and ornamented prose as flowery, purple, self-indulgent, self-conscious, clichéd, melodramatic. Drowning in its gendered excesses, metaphor is fundamentally abject.

On another occasion I attended a conference devoted to the military metaphor in medical discourse. Many of the presentations alighted on Sontag's argument and were largely united in their opposition to the military metaphor: it was at best lazy and unhelpful, and at worst violent and discriminatory, encouraging a binary view of a 'pure' self and 'monstrous' other. The room was becoming soporific in the way that conferences tend to; now that we had established we were all on the same page, we could go to sleep on it. Then came the rupture: a GP stood up and said she agreed with those who had described the military metaphor as inaccurate and unhelpful. She concurred with Sontag that the body is not a battlefield, and that immunity and infection are infinitely complex. However, in the pressurised space of a ten-minute consultation with her patients, how else was she supposed to communicate what was going on inside their bodies? The military metaphor may not be ideal, she said, but it is at least accessible, and one she can be sure that all her patients, regardless of their backgrounds, will grasp. If we do away with the military metaphor, what alternatives do we have?

Though I would be quick to interject that limits on what can be communicated within the ever-shrinking space of a GP consultation are the fault of twelve years of austerity, rather than a failure of imagination, the GP's questions were a salutary reminder to me that ubiquity is graced by commonality; it constitutes a shared linguistic space. Language is full of these pockets of affective weather, linguistic micro-environments in which we can gather and clumsily articulate. In her writing on cliché, the poet Denise Riley suggests that the cliché is not to be despised, because 'its automatic comfort is the happy exteriority of a shared language which knows itself to be a contentless but sociable turning outward toward the world.' In Riley's book the cliché is redeemed not in spite of its ubiquity but *because* of it – its contentlessness is described as one of language's blind spots, 'which are not flaws but are constitutive of it, needed for its workings.' This reparative view of language's tics reminds me that metaphors are our shared condition.

The preface to *Illness as Metaphor* famously begins by describing illness as 'the night-side of life', stating that 'Everyone who is born holds dual citizenship, in the kingdom of the well and in the kingdom of the sick.' By surveying the two kingdoms, Sontag prefaces her book with the same spatial or territorial metaphor she later lambasts with regard to cancer as a 'pathology of space', a disease which spreads or proliferates. The first time I read this, the parody was lost on me. It was only on subsequent readings that I began to fully register the preface's impasto-like texture, thick with layers of figuration. The most truthful way of regarding illness and the healthiest way to be ill, Sontag writes, is to resist metaphorical thinking. So what does it mean that her argument is mired in the figurative substance from which it seeks liberation?

In *AIDS and its Metaphors*, published twelve years later, Sontag describes her use of metaphor in the earlier book as a 'brief, hectic flourish [...] in mock exorcism of the seductiveness of metaphorical thinking.' Well, consider me seduced. But in her invocation of life's night-side and its attendant shadowlands, Sontag adumbrates a space of experience that is perhaps inimical to 'purified' language: a space made of blind spots into which only blind spots can venture. Riley's embrace of the cliché's contentlessness is a talisman for reading Sontag, a reminder that it might not be possible to cast light on the shadowlands, to peel off the fripperies of metaphor and image and arrive at the kernel of truth, some essential expression finally released from the shadow of projection, irradiated by its own authenticity.

If language is riddled with pestilence in the form of metaphor and cliché, these are perhaps a linguistic parallel to the role of parasites and diseases within an ecosystem. Riley might have been speaking about disease ecology when she wrote about cliche being 'constitutive' of language, 'needed for its workings.' My ecology textbook informs me that more than half of the earth's individual organisms are themselves parasites, and, as Carl Zimmer writes, 'Every living thing has at least one parasite that lives inside or on it'. To varying extents, we are all born under the sign of parasitism.

At the same conference, the one about the military metaphor, I got speaking to a medical anthropologist, C—, working on bubonic plague. Plague is the jewel in the crown of zoonotic diseases – infamously killing between thirty and sixty percent of Europe's population during the Black Death of the fourteenth century. In most people's imaginations plague appears – as it used to in mine – as a purely medieval phenomenon. But plague, in all its bubonic and pneumonic morbidities, is still endemic in many parts of the world, including Madagascar, where the disease flares up each year with seasonal regularity. In Mongolia, the infectious *Yersinia pestis* bacterium is carried by marmots, and occasionally transferred to humans through contact.

If caught early enough plague is treatable with antibiotics; in practice, treatment is often compounded by the unlikelihood of diagnosis. In its initial stages plague manifests as a series of flu-like symptoms – you would have to have some idea of your exposure to plague in order to seek treatment for it. The levels of human death by plague in Mongolia are relatively low, as they are around the world. But they are still there. C— explained that in Mongolia this needed to remain the case, because eradicating plague would lead to an overpopulation of marmots, whose appetite for grasses and herbs would eventually lead to the desertification of the landscape, ostensibly posing a much greater threat to human life than the currently low levels of endemic plague.

I would go on to hear echoes of C—'s unlikely defence of plague in conversations with others well-versed in disease ecology. There appeared to be an ongoing adjustment in epidemiological standpoints; it was frequently mentioned that the 1980 eradication of smallpox (to date, the only infectious disease affecting humans that has been globally eradicated) set a false precedent. The language was shifting from eradication to tolerance, resistance and adaptation. But understanding what is ecologically sound doesn't necessarily make it any more palatable. Masochistic tendencies aside, my response to my body's harbouring of parasites isn't to lie back and feel graced by their presence, grateful to be a node in the web of ecological diversity.

Listening to C— talk, I couldn't help but be disturbed by a Malthusian undercurrent, one that might be all too easily co-optable by those who consider humanity – or some of it – a scourge on the planet, righteously culled by war and epidemics. Like Pasteur, the eighteenth-century economist and reverend Thomas Robert Malthus was not a fan of the poor. He believed that while populations can increase exponentially, resources will never be able to keep up, eventually causing war or famine which bring about a natural curb to overpopulation.

Long before the 'humans are the virus' contingent of Covid-19, I started wrestling with the neo-Malthusian outlook at sixth form, provoked by my French teacher, Monsieur S—. Our lessons took place on the highest floor of the school's tower block, overlooking the housing estates that dribbled down to the sea. Sometimes on a slow

Tuesday afternoon, our attention flagging on subordinate conjunctions, Monsieur S— would stride over to the window and cast his arm wide over the view. 'Humanity has a natural way of culling itself when population numbers get too high,' he would say, a daring gleam in his eyes. 'Don't you think we should allow diseases to take their natural course, and kill off the weaker members of the human race?'

The room began to twitch. It was a run-down, perpetually failing school, and many of us were born and raised on the estates on which he was wishing symbolic death. Some students, inured to his provocations, would roll their eyes, while others couldn't resist the bait. He wanted us to articulate and debate, but we were tired, reactive. More often than not, his devil's advocacy backfired and reduced us to silence. I've since come to wonder if the grammar we couldn't focus on in his classes might have been better psychological preparation for our current ecological quandaries. Those who make Malthusian claims don't usually include themselves in the equation; populations are reduced to abstract quantities rather than actual, situated lives. At the military metaphor conference someone suggested that the main obstacle is the centrality of nouns in the English language, and I wonder if Malthus had the same problem: grammar allows us to speak of 'populations' as if we are not part of them.

Before the pandemic I found myself at something of an impasse. Trying to resist the military metaphor in relation to pathogenic life became a circuitous exercise in disavowal and abstinence. It felt like the blasphemous thought-loops I experienced as a child made too aware of sin; I feared the tendency of my thoughts to stray to where they shouldn't, and so that was where they inevitably lived. Perhaps this is why, as an adult, disavowal leaves me with dangling threads. Excising a bad metaphor feels similar, conceptually, to excising a bad bug. Even though I may agree with the usefulness, and often necessity, of doing so, I'm more interested in what happens afterwards – how language might rearrange itself differently, the eradication's effect on the balance of a wider ecology.

I was coming to understand that life isn't a binary of sickness and health, good bugs and bad, and that humans aren't so human after all. Since humans contain less than 50 per cent animal cells, biologist Lynn Margulis has suggested we are more accurately a 'holobiont', an assemblage of host organism and its microbial, bacterial and viral partners. Stefan Helmreich similarly proposes replacing *Homo sapiens* with *Homo microbis*. From biologist Margaret McFall-Ngai I learned that pathogenicity is a set of behaviours that species can move in and out of, rather than a fixed biological state of being. The behaviour and communication of pathogenic microbes, she suggests, is fundamentally the same as that of mutualistic or commensal species; the latter can become the former through 'fitness peaks' and changes in the host environment. The human gut yeast *Candida albicans*, for example, is not conventionally understood as a pathogen, but if the gut flora are imbalanced it becomes opportunistic and colonial, blooming into an infection as a pathogen would.

I wanted metaphors – however atmospheric or diffuse – that would do justice to this understanding of latent pathogenicity, of antagonistic attachments, of what geographer Steve Hinchcliffe has called 'life [...] threatened by its very own liveliness'. The now-ubiquitous genre of nature writing and its healing, revelatory encounters with an enchanting natural world wouldn't cut it. It is not always easy to be on the side of life; some relationships on earth remain irreducibly violent for one party involved.

At some point the words 'love' and 'bug' copulated in my brain, and, true to their conjoined namesake – the 'lovebug' march fly, *Plecia nearctica* – were reluctant to separate. Considering relationships which – even while harmonious – were underscored by latent conflicts and antagonisms did not make me think of wars and battles, as per the military metaphor; they made me think of love. I won't offer a totalising definition of love as I use it in this book; like the Erotes of Aphrodite's retinue, the loves of *Lovebug* assume different forms, from romantic clichés and familial attachments, to mythological figures and the ecstasies of mystical union. What connects them is an understanding of love as a force which can both make and unmake the experience of self; love can both destroy the lover, and set them on new alignments that would be impossible to reach alone, should their boundaries remain intact.

Lovebug contains some personal abjections but I think it is most personal in its method – an attempt to think through, and stay with, irreconcilability. In writing it I consulted art, literature and psychoanalysis – the things I consult in order to rewild the epistemic ecocide of my upbringing. I often felt that I was playing an elaborate game of conceptual catch-up with people who I imagined had been raised with a capacity for contradiction and ambivalence in the way that some people are fortunate to be raised with progressive politics or good dental hygiene. Realising I might not ever catch up – would never be qualified for my subject – became important; I want to insist, following Lisa Robertson, 'that a poet is an amateur, which is to say a beginner, who works only in the company of language's ghosts.' *Lovebug* was written in the company of many ghosts, some friendly and some not so, including science, medical terminology, memory, failure, textbooks, class, institutions, funding, relationships, a pandemic, unproductivity, a microscope and doubt. I never get tired of being reminded that the amateur is 'someone who loves', someone whose claim to knowledge is necessarily partial, vernacular and localised. It's also an obsession, or a fleeting crush. Any object can become a love object, after all, and for a while the preoccupations of this book were mine.

Writing *Lovebug* also became a way to understand the way in which I was being recomposed by a then new, and now chronic infection. This book is not an illness memoir, but neither is it written from the fabled shores of good health. I recently learned that my infection's tenacity is likely enabled by my seemingly unrelated connective tissue disorder. Infections can wax and wane, but no amount of antibiotics can change the DNA of my tissue; perhaps it's this predisposition that keeps my fantasy of getting better for good – of winning the war – in check. On bad days I still crave this elusive victory, a happy-ever-after of health; on better days I perceive chronic pain as analogous to life, in its sometimes unbearable, sometimes bearably quotidian little anguishes of being. Health is perhaps as flighty and mercurial a force as desire – something I am compelled by and moved to pursue but can never fully arrive at or possess, except through what would amount to a kind of death.

Writing as someone who is known to have written poems means that my prose is at risk of being described as lyrical or poetic; I neither encourage nor resent this, but in the context of this particular book it seems worthwhile to preface any lyricism with the somewhat violent multispecies origins of its namesake. The first lyre is said to have been crafted by Hermes upon leaving the cave he was born in and encountering a tortoise, which gave him an idea. He killed the tortoise and cut it out of its shell, using the empty carapace, along with reeds, sheep guts and an ox hide, to make a seven-stringed lyre. In no time at all, he was singing beautiful songs of love.

Love Kills People

Wherefore, sweetheart? What's your metaphor?

– *Twelfth Night*

When was love first described as a sickness? When did the body in love begin to be likened to one battling an infection? The parallels are self-evident: love moves in and colonises all available space, reorienting everything to itself. It weaves a gauze beneath the skin of its subjects, a mesh through which all experience is filtered for the duration of the infection: healthy cells become love cells, resources are corralled and redirected, the body's tangible and intangible energies are reconstrued as a living shrine to the beloved. In most European languages the word for love is said to derive from the ancient Nostratic word *luba*, meaning thirst. Love's appetite ends either with privation for the love-starved, or else a gorging, a devouring of its objects. A consumption.

Infect me with your love / Nurse me into sickness / Nurse me back to health – when Matt Johnson of The The wrote these lyrics he was tapping into the familiar metaphor of love's virality. Ingeborg Bachmann's novel *Malina* figures love as a life-threatening virus, a sinister ploy of patriarchy and fascism to ensnare, and finally erase, women. In *Twelfth Night*, romantic love is figured as a contagion that transfers between and consumes its victims: the noble duke Orsino becomes atrabilious under its influence, and lady Olivia – who fails to return Orsino's affections – dubs it 'the plague'. Upon first seeing her, Orsino believes Olivia's beauty has 'purged the air of pestilence', but he appears to internalise love's pestilential appetite in the process. The lover is the one who gets sick by eating the apple of their own eye, bugs and all.

If love is a disease, is its nature contagious? Or does it derive from and degenerate within the lover, like a malignant growth, or theories of spontaneous generation, in which the agents of decomposition – worms, maggots – grow spontaneously from inert matter? For Donna Haraway, love is a 'nasty developmental infection' signified in the flesh. For Sappho it is a quasi-malevolent force which intrudes from the outside as 'neither inhabitant nor ally of the desirer. Foreign to her will, it forces itself irresistibly upon her from without. Eros is an enemy.'

Drawing on conclusions in Greek medicine about love as a disease that may culminate in death, the tenth-century Persian physician Avicenna declared falling in love a 'disease of sadness'. By his count, physiological symptoms of the lovesick include irregular pulse, dry eyes (except when crying), rapid fluctuations in mood, malnourishment, increased respiration and lack of sleep. Avicenna concluded that medicine could offer little help for lovesickness, which must be allowed to run its course. In Veronica Forrest-Thomson's poem 'The Garden of Proserpine', it is the police rather than the medical establishment that does little to stem love's murderous impulses:

> Love kills people and the police can't do anything to stop it.
> Love will:
> > ravage your beauty
> > disrupt your career
> > break up your friendships
> > squander your energy
> > spend every last drop of your self-possession
> Even supposing you had such qualities to start with.

The poem understands love as a deadly force that overpowers all attempts to contain it. Although 'The Garden of Proserpine' resists depicting love as a character or figure, its parasitic nature and dependency on human hosts is implicit: 'It won't leave us alone for a minute. / For without us it wouldn't exist.' It's a poem attuned to love's double bind, to the humiliating inevitability of being made a fool of by love, and the greater folly that lies in trying to resist this fate, 'To pretend a stoic indifference'.

Love's lesson, at least according to this poem, is that vulnerability might be our undoing. Just as we all react differently to love's arrows, each of us responds differently to new, potentially harmful pathogens. Even so-called 'healthy' human bodies harbour communities of latent pathogens, which may be activated through contact with external agents. Love may enter me like a microbial infection, but like these infections, it is never entirely one-way – it activates and is activated by the latent pathogens of love. When Freud sought a metaphor for cathexis – the means by which the ego attaches libido to its love-objects, and then retracts back into itself – he chose the shapeshifting (and, I would add, potentially pathogenic) amoeba. But this analogy already makes me uneasy, a troubled straying between sickness and love, infection and affection. In the work of Denise Riley I discover that 'affection' is an archaic term for disease; language conspires to bind infection and affection too close for comfort.

⁂

When I tell people I am trying to think about infection through the lens of intimacy, it's assumed that I am talking about what can be transmitted sexually. I reply that sometimes I am, but that the sex in question might not be between humans but between parasites that mate or reproduce inside the body of a host. That in other cases the sex is not sex as we think of it but the asexual multiplication of bacteria, blooming into infection. I let down the boundaries of my body to let the outside in – once inside, the outside I contain may destroy me.

In Hervé Guibert's novel *Paradise*, for instance, written six months before the author's death from an AIDS-related illness, the young narrator and his lover Jayne travel from Martinique to Mali, suffering from a number of infections. The infections seemed to be connected to the intensity of their sexual entanglement: the narrator has brought a gun 'for protection', and Jayne begs him to fuck her cunt with it. He concedes, and it becomes a regular fixture of their fucking. Later, they have sex on top of a glass fish tank, and at the point of climax no longer know if they 'are fish or gods'. Their infections worsen, and towards the end of the novel the narrator realises that 'I re-contaminated her. We'll be passing on these *chlamydiae* to one another forever more, until the final extinction of our hypertrophied genital organs, holed and worn into sieves'.

Equally suggestive, if subtler, is John Donne's poem 'The Flea', believed to have been written in the 1590s. It's perhaps misplaced to read this poem with any zoological realism in mind; the flea was an established symbol in Renaissance erotic poetry, in which the speaker of the poem typically describes envying the flea for the liberties it takes with the flesh of his beloved. Donne's poem sits comfortably in this tradition. The flea acts as an extended metaphor throughout the poem: first it sucked the blood of the speaker, and now it sucks that of his beloved. The speaker points out that, even though his beloved refuses to be intimate with him, the bloodsucking flea has already brought their bodies together:

> It sucked me first, and now sucks thee,
> And in this flea our two bloods mingled be

In a Renaissance context, this mingled blood represents consummation and marriage. The rhetorical subtext here is the speaker begging his beloved to have sex with him; she may as well, he argues, as their flea-mingled blood has already designated the creature's body their 'marriage bed' or 'marriage temple', as if they have metaphysically fucked inside it:

> Oh stay, three lives in one flea spare,
> Where we almost, nay more than married are.
> This flea is you and I, and this
> Our marriage bed, and marriage temple is

Creepy hook-up poem par excellence, I also can't help but read 'The Flea' as a proto-epidemiological fable, one that traces the difficult meshwork of infection and intimacy. If three lives can exist in one flea, life and liveliness is something that can move across species boundaries. If the allegorical flea in the poem is also carrying bacteria, it is more than just pseudo-conjugal blood that passes between the speaker and his beloved; like the *chlamydiae* of Guibert's novel, the flea marks an infection 'passing on [...] to one another forever more'.

Thinking about the epidemiology of this poem becomes even stranger in the context of Donne's near-death experience with a relapsing bacterial fever, thought to be typhus. Typhus is caused by *Rickettsia* bacteria which are typically transmitted to humans by fleas. Perhaps Donne was tempting death by writing allegorically about the creature that would bring him close to it. In any case, it was a closeness that also brought him insight and posterity; his famous expression 'no man is an island' was written during his recovery from the illness, appearing in *Devotions Upon Emergent Occasions, and severall steps in my Sicknes*. Published in 1624, the *Devotions* reinscribe the Renaissance belief in illness as a visitation from God, a trial from heaven, the faith of the believer tested against the frailty of the flesh. In this epidemiology of the spirit, Donne's 'island' is the drift-space between man and God, life and death, sickness and health. But there's an infectious undercurrent to Donne's writing too, I think, both in 'The Flea' and in *Death's Duell*, the final sermon he preached in St Paul's Cathedral, which describes the horrifying 'vermiculation' of being eaten alive by worms. To be made of flesh is to be vulnerable; for better, for worse, in sickness and in health – no species is an island.

⁂

The first literary biography of a bacteria was written in 1935 by bacteriologist Hans Zinsser. The subject of his book, *Rats, Lice and History: The Biography of a Bacillus* was also typhus. In its historiography of disease the book is anecdotal bliss ('Archelaus believed the putrefying spinal cords of animals and man were transformed into snakes'), but its scientific narrative only seems able to progress through a curious disavowal of both contemporary literature and psychoanalysis. Justifying his choice of literary genre, Zinsser's preface laments the way biographical writing had displaced the novel, erotic memoir, detective story, scientific literature and 'even entered the realm of the psychopathic clinic'. In addition to its scientific content the book contains critiques of Sigmund Freud, James Joyce and T. S. Eliot. Gertrude Stein's writing is parodied by Zinsser's evocation of an institutionalised woman who believes she is a baby. 'Incoherence and a bad smell,' Zinsser writes, 'don't make a Baudelaire'.

I get the sense that Zinsser would be equally dismissive of my epidemiological reading of Donne – and in fact his sneering tone comes close to embodying *Lovebug*'s superego, mocking my illicit burrowing between disciplines and amateurish adjacency to science. Just as I was beginning to consider how certain psychoanalytic concepts might be useful to think about the difficulties and contradictions of infectious life, there was Zinsser:

> We will profit by no assistance from psychoanalysis. There will be no parental influences; no Oedipus or mother complexes; no early love affairs or later infidelities; no perversions, urges or maladjustments; no inhibitions by respectability, and no frustration by suppressed desires [...] [For] the subject of our biography is a disease.

A contemporary of Freud, Zinsser was clearly unsettled by the rise of psychoanalysis throughout their lifetimes. And yet psychoanalytic thinking has a lot to offer pathology, since few other frameworks can better articulate and tolerate a coexistence with harm – both the harm done to us, and the harm we do to ourselves. The military metaphor doesn't hold up, for example, in light of a scientific discovery that found viral infection to be intrinsic to human evolution. At the turn of the millennium, an unusual gene in the human placenta, syncytin, was discovered to be responsible for a protein which fuses the placenta to the wall of the uterus. Without this structure in place, the foetus cannot draw nutrients from the person carrying it. Scientists also found that syncytin suppresses the parent's immune system so that they do not attack the foetus growing inside them as a parasite. What turned out to be unusual about syncytin is that it derives from a virus, variations of which infected the ancestors of most carnivorous mammals some eighty-five million years ago. Humans, primates, mice, rabbits, cats and dogs all have a type of syncytin in their genome. As virologist Carl Zimmer puts it, 'If not for a virus, none of us would ever be born.'

How to accommodate this conflict – that a virus is both that which can harm or kill me, and that through which I came to be alive? Humans are estimated to be about 8 per cent virus and 37 per cent bacteria. Most organisms share up to a third of their genome with other species and kingdoms. Psychoanalysis has taught me that I am not what I imagine myself to be – and this also scans at the level of phylogeny; *not so human after all*. I imagine Zinsser was affronted by Freud's insistence on psychoanalysis as a serious science (this being the cause of his rupture with Jung, and the latter's increasingly spiritual and esoteric interests), even though Freud's theory of the death drive was informed by his reading of Lamarckian evolution. In a paper unpublished during Freud's lifetime and only rediscovered in the 1980s, the link between pathology and psychopathology was made even more explicit. This paper, 'Overview of the Transference Neuroses', attempts to trace the roots of neurosis back to the ancestral trauma of living through the first Ice Age. Anxiety became inscribed in human phylogeny, Freud suggests, when the once friendly primeval world turned icy and hostile, and 'transformed itself into a mass of threatening perils.'

In the forty-seventh Annual Freud Lecture, 'To die one's own death – thinking with Sigmund Freud in a time of pandemic', Jacqueline Rose contextualises Freud's writing of this paper with the details of his own life. By 1918, the effects of World War I had plunged Freud's family into destitution and near starvation. When his favourite daughter Sophie began to die of the pandemic flu strain, he was unable to find a mode of transport to go to her. Experiencing such grief and poverty, Rose says, led Freud to hypothesise about how extreme threats to existence imprint themselves upon the psyche. Humanity's propensity for anxiety and neurosis, Freud proposed, is the inherited result of our ancestors' attempts to survive hostile environments, such as the Ice Age. It was one of Freud's closest interlocutors, Sandor Ferenczi, who pursued the biology of psychopathology in greater depth. Throughout his clinical research Ferenczi elaborated what came to be known as a theory of the 'biological unconscious', blurring the Cartesian dualism that separates body from mind, suggesting that organs and organisms are not just brute matter, but suffused and entangled with psyche.

This idea – of the body's formative response to its environment as separate from conscious thought and memory – is perhaps an easier sell for contemporary readers, since concepts of trauma and its effects on the body have been absorbed into popular culture; *The Body Keeps the Score* is an international bestseller. Citing a huge array of medical case studies, the Lacanian psychoanalyst Darian Leader, writing with philosopher David Corfield in 2007, has explored the relation between illness and the emotional life of the subject, or what Ferenczi would call soma and psyche. They survey inflammatory illnesses as responses to grief and study the phenomena of what Ferenczi called 'anniversary reactions' – the appearance or reappearance of physical symptoms on a date significant to the subject, usually a bereavement. If being able to psychically integrate what has happened to us – pain, loss, abuse, betrayal – is the goal of the so-called talking cure, Leader and Corfield suggest this integration – or disintegration – also affects how loss is inscribed in the body. 'When we are unable to think something through,' as they put it, 'the body may suffer the consequences.'

Ferenczi's insistence on approaching matter from 'the other side', i.e., from psychoanalysis, reminds me of another psychoanalytic borrowing, albeit in a different direction. In a seminar during his training year at the French Analytical Society in 1953, Jacques Lacan informed his students of the influence of geology on their nascent discipline:

My dear fellows, you wouldn't believe what you owe to geology. If it weren't for geology, how could one end up thinking that one could move, on the same level, from a recent to a much more ancient layer?

The stratigraphic layers, pressurised fossils and shifting deposits of geology afforded Lacan a rich structural model of the psyche and a way of sidestepping the arrow of chronological time. To encounter a geological form was to encounter the present suffused with prehistory, and wasn't this also true of the analysand?

⁂

In order to stop the human placenta from rejecting a foetus as a parasite, the viral protein syncytin prevents a premature splitting of self from other, and allows for the conditions of life to grow. The impulse to split off from what might harm us is at the core of the work of Austrian-British psychoanalyst Melanie Klein, in her concept of psychic positions. These positions – the paranoid-schizoid, and the depressive – emerge in early infantile, pre-verbal life in relation to the 'objects' (caregivers, environment, resources) around us; they are also positions of the ego that can reoccur throughout adulthood. For example, in a breastfeeding scenario, the child in a paranoid-schizoid position will be unable to withstand inconsistencies in nourishment and care; it will split off the 'good' breast or parent that feeds it from the 'bad' breast or parent that is sometimes in a hurry, or in pain, or for any reason fails to deliver. The long-term result of this splitting is a binary view of people or things in the world as good or bad – the latter of which become objects of hatred and are experienced as a terrifying threat to the self.

On the other hand, the depressive position is able to accommodate these discrepancies in people and objects around them, understanding the nature of inconsistency, the fact that good and bad mingle from the same fallible source. If the military metaphor correlates to a paranoid-schizoid view of microbial life, in which such life is considered indiscriminately 'bad', an enemy of the self to be eliminated, then the depressive position strikes me as fundamentally ecological in nature, an outlook that can accommodate the paradoxes and antagonisms resulting from co-existence with other species and environments. In the depressive position, the difficulties arising from indeterminacy are navigated and assimilated, rather than rejected outright as threats.

⁂

In the melding of traditions that gave rise to the phenomenon of courtly love, sickness and disease were understood as inherent properties of romantic love. Scholars such as Idris Shah trace courtly love's line of influence back to Islamic and Sufi mystical traditions, whose songs extol the unreachable divine in the way that courtly love extols the beloved. The beloved seems to experience the passion of mystical union from a depressive position: there's an understanding that the divine lover other can be loving and intimate, also jealous, angry and violent. Some medieval Christian mystics describe an explicitly physical, at times carnal, relationship with God. Catherine of Siena claims to have worn Christ's foreskin as a wedding ring; Marguerite Porete wrote of being melted by God and remade by him in 'the middle of the marrow of divine love'; and Teresa of Ávila gleefully incited believers to death, comparing their bodies to those of silkworms which are boiled alive inside their cocoons in the process of making silk:

> Die! Die as the silkworm does when it has fulfilled the office of its creation, and you will see God and be immersed in His Greatness, as the little silkworm is enveloped in its cocoon. Understand that when I say 'you will see God,' I mean in the manner described, in which He manifests Himself in this kind of union.

If divine love is a parasite, it is one whose victims often volunteer for depletion. *Alukah* is a Hebrew word for 'leech' or 'horse-leech' that appears in a riddle of King Solomon: 'The leech has two daughters – Give *and* Give!'. To interpret these two 'daughters', biblical scholars have looked to *alaqa*, the Arabic word for leech, which derives from a verb that means hanging from or adhering to. As artists Himali Singh Soin and Tyler Rai have put it, 'alaqa is a clinging kind of love, a passionate attachment, a symbiotic care or a parasitical slurping.' In Jewish mysticism and Babylonian demonology *Alukah* has been variously characterised as a vampire, a flying, lycanthropic shapeshifter and a succubus. The leech and her daughters might be said to embody the parasitical or monstrous nature of love's appetite.

In her notebooks, the philosopher Simone Weil drew

on parasitic ecology to adumbrate divine love, and its movement from microcosm to macrocosm:

> [A]s a parasite lays its eggs in an animal's flesh, God places a sperm in our soul which, when it has grown, will be his Son [...] We ought to give our soul as food for this germ. After which it will itself obtain direct nourishment from everything that formerly nourished our soul. Our soul is the egg within which this divine germ grows to a bird. As an embryo, the bird feeds on the egg; once it has grown it breaks the shell and emerges and pecks for grain. Our soul is shut off from all reality by an enclosing skin of egoism, subjectivity, and illusion; the germ of Christ, placed in our soul by God, feeds on this; when it has grown enough it breaks the soul, explodes it, and makes contact with reality. That is Love in the microcosm. Love in the macrocosm, when once its golden wings have grown, breaks the egg of the world and passes to the place beyond the sky.

For Weil, God is a parasite whose love breaks us apart, illustrating the principle of decreation – self-undoing – that pervades her thought and writing. The move from microcosm to macrocosm necessitates a puncturing of the self that leads to divine communion, 'the place beyond the sky'. Mystical union is a spiritual infection in which the boundaries between self and other are broken down and dissolved; but rather than a fearful schizoid reaction to this invasion, the loss of bodily coherence is experienced as a sublime intimacy with God.

I was going to write that my interest in mystical union comes from lived experience, rather than academic or aesthetic interest, but even the attempt to do so seems to place it at arm's length. Mystical experience doesn't belong in sentences, or at least not ones like these. I can say that I grew up hearing people speak in tongues, convulse with the holy spirit, heal the sick and cast out demons. I could write that I no longer believe in God and perhaps never did, but that as a teenager I experienced states of trance-like worship in which an invisible wave knocked me from my feet to the carpeted church floor, or that I once felt a warm sweet oil being poured over my head. I had faith in these gestures, which seemed unthought and therefore true, beyond my control. Speaking in tongues was a different matter: it was too close to language, which for me was close to thought. My grandmother, who I lived alone with, was skilled in glossolalia and the receiving of visions. She told me there was nothing wrong with 'practising' – you simply open your mouth and babble like a baby; it doesn't matter if you dribble a bit, since God understands it all as praise. Speaking in tongues was an individualised language of worship that, like any language acquisition, could be honed. When someone's tongue was sufficiently developed, God would use it to deliver messages to the wider congregation by granting someone else the 'interpretation' of a given tongue.

Oh for the mouth to be a conduit: there were years I so desperately wanted it. I didn't want to practise, I wanted to feel possessed by an unknown language. I broke up with the church at seventeen, and a year later my grandmother broke up with our living arrangement: she was moving to a church five hundred miles away, and since I had reached the legal age of adulthood, her duty was done. I became a lodger and did an art foundation course. I replaced my bible with a different kind of piety – the writings of the Marquis de Sade and Georges Bataille and, a few years later, Simone Weil.

I didn't stay with my grandmother again aside from a few weeks in my early twenties when I returned from a study-abroad year in Finland. It had not gone well while I was away. All my childhood demons had caught up with me: I was sick and skeletal and not sleeping, quarantined in suffering and unshakeably convinced of my own evil. That I went to her was a sign of both my desperation, since there was no one else, and of my will to self-destruct. I remember lying down with my face pressed into her scratchy carpet, shivering as she kneeled over me, her hands moving up and down my back. She told me that she knew I had been suffering because God had granted her a vision while I was away. There was a demon affixed to me, she said, twisted around my spinal cord like a snake. She traced spirals down my back as if to determine its outline. The demon's grip was tight, she concluded, but she would do her best to exorcise it.

I can write about these things, but what I feel I am writing into is the gap between what happened and the words that inadequately fix it here: I can mimic the tone of memoir and give the outline of an account, but, as my grandmother taught me, the outline isn't the demon. The rejection of closure has a mystical counterpart, which is negative theology or 'apophasis'. A language of unknowing, it proceeds from the understanding that God is beyond human comprehension and so cannot be accurately described; to say that God is unnameable we have already contradicted ourselves by naming them 'God'. Instead, the believer might get closer to the divine through a devotional mode that takes contradiction as its subject, acknowledging that whatever is said about God must then be unsaid. It is the tension between saying and unsaying that gets closest to the unnameable. It's a language of doubt, paradox and not-knowing. I am still talking about pathogens, since we are in the realm of corporeal possession. Love may not be a literal infection and infection may not be a literal love, but isn't there something in the tension formed by moving between these two positions – a glimpse of some gap-thing, in the slippage between demon and outline?

The Bite

> *"I am afraid of being bitten" or "I am afraid of biting"?*
>
> – Julia Kristeva

Love nips or bites, the skin is bruised or is broken. I had one dog as a child, a border collie named Poppy. Poppy was what is universally understood as a good dog. She would drop anything that happened to be in her mouth whenever my grandmother ordered it. To exhibit Poppy's superlative goodness to others I would show off by putting my hand between her teeth while she panted and raised her eyebrows – cute. See, she's a good dog, I'd say, she'd never bite.

The good dog takes my hand in her mouth but does not bite down. I feel the chalky points of her teeth against my skin, on which signs are indented when I retract my hand, love-words glistening with saliva. The good dog leaves a dent but not a wound, a dent which nods towards the possibility of puncture and says *I am not the wound, but*

I could be. I am exhilarated by this proximity to the wound that isn't, both disappointed and relieved by the encounter from which I emerge intact.

Poppy was preceded by a dog I barely remember except for her grey shagginess, Nell, who was already owned by my grandmother when I went to live with her in the months before I turned two. I don't remember what came before this but I have been told it was just me and my mum in a rented flat, no dog. She was twenty, had been a teenager under Thatcher, and was not immune to the endemic shaming of young single mothers that typified the era. There was little in the culture that kept her from thinking she was a bad single mum, that I would be better off living with her mother while she studied to get a better job to support us. The two of us were living off £67 a week and how long could that last? My grandmother had a garden, a mortgage, a dog and at that point a husband; the Oedipal dream was in reach.

But the house and the garden already had a child, even if that child was an elderly dog. Nell did not take kindly to being usurped by me, the new baby. My grandmother claims to have watched through the kitchen window as Nell went for me at the bottom of the garden, where she must have been biding her time, thinking we were out of view. I don't remember the incident, or being upset by it; there was a bite on my ear but nothing serious. But Nell, my grandmother decided, had to go, lest she act on her instinct to maul me to death. Nell was put down, and we forgot about her. Within a few months there was Poppy.

⌘

In *A Thousand Plateaus*, Deleuze and Guattari distinguish three kinds of animal. First, the 'Oedipal animals', sentimental animals or pets ushered into the familial scene, ripe for regression and projection. Second are the anonymous 'State animals', those that exist abstractly as species belonging to lists and data. Lastly are 'demonic animals', a pack or swarm that exist in multiplicity, those that form the minatory backbone of myth. But these categories are not absolute or discrete. Deleuze and Guattari write, 'There is always the possibility that a given animal, a louse, a cheetah or an elephant, will be treated as a pet, my little beast.'

Donna Haraway's *The Companion Species Manifesto* concurs. Haraway is suspicious of those who claim dogs' capacity for 'unconditional love', writing that treating dogs like surrogate children or relatives is 'abusive – to dogs and to humans.' Rather than the adulation of Oedipal animals, Haraway advocates an interspecies love of relating in specific difference, recognising the ways that humans and dogs have co-evolved and share aspects of our microbial profiles. Speaking of her dog Cayenne Pepper, Haraway writes:

> Ms Cayenne Pepper continues to colonize all my cells – a sure case of what the biologist Lynn Margulis calls symbiogenesis. I bet if you checked our DNA, you'd find some potent transfections between us. Her saliva must have the viral vectors [...] We are, constitutively, companion species.

This molecular intimacy is one that accommodates both loving companionship and microbial transfer, acknowledging that many infectious diseases affecting humans arose from our proximity to domesticated animals. In the thousands of years of humans' close-knit relationship with domesticated animals, microbes have had time to bloom and make the connection. This is why measles is so closely related to rinderpest, a disease of cattle, and pathogens related to those causing tuberculosis, smallpox, flu, pertussis and falciparum malaria can be found in cows, pigs, ducks, dogs and chickens. Given this co-constitution of pathogenic life, I know in theory that I might become infected by the bite of any one of the animals Deleuze and Guattari describe. Yet somehow, it is the bite from an Oedipal animal that hurts the most, the one that I don't see coming.

This was also the case for novelist Joy Williams, as described in an essay about her dog Hawk. Hawk is a black German shepherd, nine years old, and devoted to Williams as she is to him. Hawk is 'my sweetie pie, my honey, my handsome boy, my love', Williams writes. She describes an idyllic day where the two of them picnic on a beach with a little fire and a beautiful sunset. But 'on the following day,' Williams writes, 'he would attack me as though he wanted to kill me.' When Williams attempts to leave Hawk in a kennel for the night (only one night, but Hawk isn't to know that), nine years of domestic companionship is hotwired by some chemical alteration in Hawk's brain. In the space of a few seconds, the affectionate nip is overridden by the atavistic bite, an impulse scaling centuries of domestication: 'I thought he had bitten off my nipple. I thought that when I took off my blouse and bra, the nipple would fall out like a diseased hibiscus bud, like the eraser on a pencil.' Hawk hasn't bitten off her nipple, but he has left deep puncture marks in his owner's hand and breast, the latter 'bruised black'.

When the good dog bites I might say it is because it is infected. The infection – whether pathogen, latent atavism or psychological lapse – is the 'bad' intruding on the dog, a splitting off which allows me to maintain the good dog's integral goodness. I could try to calm the biter and empathise with its bite-words, bywords for fear or threat, but by that time my blood is flowing and my skin hangs in half-chewed ribbons. I struggle to accommodate the shattered illusion of my perfect Oedipal animal, to acknowledge the sweet dog as a complex Kleinian entity, the good and bad in one ambivalent, nonhuman other.

Joy Williams is surprised that the bite on her breast and her hands needs medical attention; the surgeon at the hospital tells her that, because it was caused by a dog's bite, 'the situation is actually life-threatening.' The bite could work its way down to her bone and infect it. My little beast could never, but it did.

When I first read Williams's essay it wasn't my own (insignificant) dog bite that came to mind, but something murkier, more in tune with the erotic undertones of its language. The term Oedipal animal, after all, doesn't specify which familial relation the animal will assume. For Williams, at least at the level of language, Hawk is somewhere between a child and a lover: *sweetie pie, honey, handsome boy, love.* Then, in a love bite gone wrong, it's her mother/lover breast that gets bitten, triggering the nightmarish fantasy of losing a nipple. Is this lactation anxiety, or interspecies melancholia? The whole scene is like a Kleinian case study written by Bataille – the bad breast punished by the child's erotic bite, the abject climax of a severed nipple, falling to the ground like a flower bud.

Mutt was a dog owned by the family of a childhood friend. This friend, whose name and biographical details have obviously been altered here, lived in a rundown house at the edge of the city. Let's call him Lee. My friendship with Lee lasted for the length of time I was at his school, roughly between the ages of seven and nine. The affinity was simple: we were both contrary and liked climbing trees. Our other friend was a girl who claimed to have once melted her barbies down to goo in a frying pan.

We spent a lot of time at Lee's since his parents were laid-back and his house was big. It was messy in a way that I loved – the polar opposite of my grandmother's – and noisy with animals and a revolving cast of Lee's large extended family. His house backed onto the woods which is where we spent most of our time making dens and potions. Mutt always came with us 'for protection'. Mutt was big and golden and faithful. Mutt wasn't the dog's real name either.

One day I was over at Lee's and his cousin, a few years older, asked if we wanted to see a secret. She ushered us into her room where Mutt was already installed. She closed the door behind us and lay down on the floor. Bent knees, nightie hiked up to hips. Her hands beckoned, and then gently guided Mutt's head into position. It's always the details that petrify such memories: her nightie was a long grey T-shirt that used to be white, emblazoned with the blue-nosed Me to You® teddy bear. The slapping sound of Mutt's tongue. There's not a point to be made from this, except: Mutt became a true Oedipal animal. We'd all had some form of bungled sex-ed by then, but in that moment I don't know if any of us were thinking in terms of capital-s sex. Lee and I never spoke about it; I changed schools soon after and never saw him again. I used to recollect the memory as a rare observation of pleasure without shame, since the cousin's only narration of the act was that it felt good. But then, if there was no shame, it wouldn't have been introduced to us as a secret. Unless, of course – as I suspect may have been the case – being witnessed, even shamed, was bundled up in the pleasure.

⌘

Kafka's short story 'A Country Doctor' is full of bites: human and nonhuman, one that stands in for sexual violence, one that seems to augur all the world's wound. It is a deep winter night, and the country doctor faces a dilemma. His horse is dead, and he is frantically searching for another to ride the ten miles to his patient's sickbed. Since the doctor doesn't have a moment to lose he takes the help of a man he finds living in a pigsty. When the doctor's servant girl, Rose, goes to help him saddle his horse, the man from the pigsty grabs her and makes his mark: 'on her cheek stood out in red the marks of two rows of teeth.' The doctor calls the man a brute and tells him to accompany him on the ride. He has read between the lines on Rose's cheek; he knows exactly what will happen to her if he leaves her with this man. But the man refuses and a silent transaction is struck: if the doctor wants to borrow his horse, he must leave Rose behind.

When the doctor reaches the sickroom, the air is unbearable – perhaps from the wound in his patient's side he discovers too late, or perhaps from the miasma of his guilt. It is as though Rose's bitemarks have blossomed into his patient's 'Rose-red' wound, which on closer inspection, the doctor sees is infected: 'Worms, as thick and long as my little finger, themselves rose-red and blood-spotted as well, were wriggling from their fastness in the interior of the wound toward the light, with small white heads and many little legs.' He concludes that the boy is past helping, and in the end all he can administer is a lie; 'your wound is not so bad', he tells him, before fleeing the scene. Buffeted between wounds and bookended by man-bitten and worm-bitten flesh, the doctor falls through the cold night into despair.

The Rose of Kafka's story reminds me of the worm-bitten addressee in Blake's poem 'The Sick Rose':

O Rose thou art sick.
The invisible worm,
That flies in the night
In the howling storm:

Has found out thy bed
Of crimson joy:
And his dark secret love
Does thy life destroy.

Read into Kafka's story, Blake's 'invisible worm' is a shapeshifter – at once the rapist from the pigsty, the literal worms sucking the life from his patient, and finally, the doctor himself, who consigns both Rose and his patient to their fates. The story would seem to suggest that none of us is immune to the 'dark secret love' of the worm – neither immune from being its victim or its enabler.

Reading this story over and over makes me realise it's a story about reading, except the texts are made of flesh instead of words. The doctor reads the marks on Rose's face, and then he reads his patient's wound, divining bad omens from both. In ancient Rome, a reader of viscera was a haruspex, a religious official tasked with identifying omens in the entrails of sacrificial animals. In Kafka's story, Rose is nothing if not sacrificial – her bite is the bad omen that only the haruspex doctor can read. But maybe the doctor is a bad reader, or simply – in a way I recognise since I am often guilty of the same thing – allows his reading to become overinterpreted. I agree with his reading of Rose's face – the mark on her cheek may be just a nip, but we all know when a nip is a bite waiting to happen. The patient's wound seems different though – is it really that bad, or are we seeing it through the distortion of the doctor's guilt? Are the worms really going to cause the patient's death, or are they in fact maggots, clearing away infected tissue so that treatment may take place? In the wake of his flight from Rose, I get the sense that his patient's wound is already overdetermined; by the time we get there it is too late to read what it says.

But perhaps I am getting ahead of myself: sometimes a nip is just a nip, as any good dog will tell you. I am often, after all, a bad reader of the world, and have to keep my haruspicy impulses in check. The anthropologist Gregory Bateson has a famous essay about being able to distinguish a nip from a bite. The idea came to him while watching two monkeys playfight at the zoo. The playfight may be made up of the same gestures and signals as real combat, but crucially differs: the playfight is a *trompe l'oeil*, a trick of the eyes and teeth to dress nips in the guise of bites. Our ability to 'read' the nips amounts to both seeing and dismissing the mirage of the bites. For Bateson this is a crucial step in the evolution of communication. Without it, there would be no metaphor, fiction or fantasy. The map would be conflated with the territory, dreams with reality. The nip would be taken as a bite.

Bateson's thinking takes the form of what microbiologist Lynn Margulis would call a 'big like us' focus on humans and mammals. The 'big like us' mindset leads us to forget what is out of sight – the microbes that pass invisibly through encounters between different animals and species. Whether playfight or real fight, the monkeys Bateson observed at the zoo would also have been subject to each other's bacterial and viral partners. The nip and bite paradigm doesn't run all the way down, there is no way of telling pathogens, *No, it was only play, I was only joking*... So while Bateson has a point in stating that 'The word, "cat" has no fur and cannot scratch', the word 'virus' might really infect us, if I am host to a particular strain, and some of those droplets spray your way as I enunciate.

Before her death in 2011, Lynn Margulis was rumoured to be working on a book about Emily Dickinson. She lived next door to the Dickinson house in Amherst, Massachusetts, and many of her friends, colleagues and students have since recounted how Lynn enjoyed committing the poet's verse to memory. 'Emily Dickinson talks to me all the time,' Margulis said shortly before her death. 'She exposes pretensions. She is a botanist. She is my favourite poet.' It makes sense that a poet who devoted unparalleled attention to life's minutiae would appeal to a microbiologist immersed in worlds unseeable to the naked eye. For Margulis, Dickinson's poetry is akin to biosemiotics, which she described as being 'alive to the cycles and mysteries of the natural world, bodily sensation, and the significance of symbols, signifiers and phrases'.

The nebulous field of biosemiotics is held together by the notion that humans are not the only sign-making (significant) species – that everything, even the most minute form of life, has a language. In this schema, human language is just one manifestation of a wider semiotic flourishing. Human culture, according to semiotician Wendy Wheeler, is in fact 'the form that the semiotic nature of biological evolution takes with the advent of *Homo sapiens*, which is also the advent of articulate language and abstract conceptual thought.' Viewed in this light, language in Dickinson's poetry undergoes a biosemiotic slippage between nature and culture, subject and object. Written language itself is figured a site of material contagion:

> Infection in the sentence breeds
> We may inhale despair
> At centuries of distance
> From the malaria –

At face-value this poem may appear to be guilty, in Bateson's terms, of confusing metaphor and reality – immaterial language cannot materially infect us. But in opening a space between the two it allows for illicit crossings – like the viral droplets spread by uttering the word 'virus', or the myriad microbial infections breeding in a three-hundred-year-old copy of Ovid's *Metamorphoses*. Cultured by bio-artist Sarah Craske, the book wears its microbial profile on its sleeves, displaying an archive of entanglement with its readers and environments. *Every contact leaves a trace.* Craske proposes a fictional discipline of 'biological hermeneutics', in which books might be read for their material and biological content alongside (or in spite of) their textual content. The microbiome of the book – partially composed of its deceased owners and readers – gracefully animates Ovid's observation that 'everything changes; nothing perishes'.

I encounter something similar in Lisa Robertson's essay on reading translations of Lucretius in the British Library. In one manuscript, she locates a flaw in the vellum – likely an animal wound – surrounded by a yonic doodle. For Robertson, this animated absence becomes illustrative of the 'readerly site'. The reader falls into the mesh of the book, enthralled, yet 'From the point of view of the world, the site of [...] capture remains invisible.'

Robertson's positioning of reading as a 'pact' or 'capture' shares a tension between readerly passivity and activity that Dickinson's poetry sustains. In figuring language as a contagion that might really 'infect' us, the poem draws on theories of miasma and contagion that persisted throughout Dickinson's lifetime. This places it closer to the views of Galen and Hippocrates than to the germ theory that gained prominence in the years following the poet's death. While Dickinson was alive, the United States was rocked by epidemics of cholera, yellow fever, influenza, typhus and malaria. Tuberculosis was a near-constant presence, greatly feared by Dickinson's father as it ran in the family. He suspected his young daughter of consumption as she often had trouble with her lungs, suffering chronic bouts of coughing and weight loss.

When yellow fever ravaged the Southern states, quarantines were put in place and explanations for the 'noxious agent' were sought far and wide; it came to be believed that cotton itself was the carrier. When Dickinson was eleven years old, Florida had lost so much of its population to the virus that it considered postponing its admission as a state. African Americans were accused of being the cause because they were partially resistant to the virus, and later, the mass influx of Irish immigrants came to be blamed, leading to murderous discrimination at the peak of an outbreak in 1855, when a mob burned the Irish ghetto in Norfolk, Virginia.

This is one constellation of contexts in which to read the 'infection' of Dickinson's poem. Another is its play on the passivity of reading. In the poem, reading actively exposes the reader to its possibly dangerous contents, recalling medieval theories of vision – extramission and intromission – and their infectious consequences.

Extramission involved the eye sending out rays in order to see, whereas intromission involved objects emitting rays of light that reach into our eyes and allow us to see them. In the medieval world, seeing was likened to ingesting; the theory of intromission was particularly dangerous for pregnant women, as it was believed that whatever a woman looked at while pregnant, she would give birth to in likeness.

A twelfth-century bestiary advised women 'not to look any of the very disgusting animals in the face – like dog-headed apes or monkeys – lest they should give birth to children similar in appearance'. The eyes were the organ of contagion, and women were believed to be particularly vulnerable. Like animals in the act of 'venery', women were susceptible to images translated 'from outside inward [...] fertilized by the imaginary figure, [which] transform the apparition into a real quality.' Any physical defects in birth could then be neatly blamed on mothers, who should have taken more care with who or what fell under their gaze.

The vagaries of intromission seem to have lingered through to Dickinson's day as a kind of corporeal suggestibility. In her poems and letters, vision and its lack render the onlooker both privy to poetic insight, and vulnerable to what they see. When her brother Austin fell ill in the winter of 1851, Dickinson wrote to him with her advice for recovery ('warmth and rest, cold water and care'), wishes for good health, and a number of vivid descriptions of pain and illness, which she stops herself short of continuing 'lest I harm my patient with too much conversation on sickness and pain'. Dickinson fears that through reading her elaborations on pain, Austin's illness may become augmented. This is the underside of her belief in 'the balsam word' that provides a healing alternative or complement to medical treatment. Her poems do not provide cures or diagnoses, but rather they are 'guests' to keep pain company.

Dickinson's prevailing struggle with illness was ocular in nature: she suffered from eye pains and light sensitivity, and often spent long periods in a darkened room, forbidden from reading and permitted to write only sparingly in pencil ('bereft of Book and Thought, by the Doctor's reproof'). It is commonly speculated that she suffered from iritis, an inflammation of the irises. Although Dickinson was well-informed and curious about anatomy and medicine, she was also suspicious of the latter's cure-all approach and wrote about the opportunities for growth offered by pain and loss. 'Before I got my eye out' begins by mourning the loss of vision, but soon diverges into the virtues of sightlessness which afford an inner vision or knowledge. The glut of being able to see everything, on the other hand, 'would strike me dead'.

Subverting the dominance of vision makes room for alternative forms of knowing. I agree with Margulis: Dickinson's poetry pre-empts a biosemiotic view of the world, making space for what we cannot immediately see or apprehend. Composed of membranes, our bodies are permeable to nonhuman others, just as our minds are porous to words 'At centuries of distance / From the malaria –'. Dickinson may not have been colonised by tubercular mycobacteria or bitten by a malaria-carrying mosquito, but her poetics assist us in remembering that the horizon doesn't stop at the limits of the visible, with organisms that are also 'big like us'. To be infected is also to be *inflected*, biosemiotically contiguous with nonhuman life, vulnerable to morphological flux.

⌘

The microorganism that causes malaria was discovered shortly before Dickinson's death in 1886, and the discovery of the mosquito vector followed a decade later in 1897. It was a disease well-known to the ancient world; in Ayurvedic traditions it was understood as the 'king' of diseases, caused by Shiva's anger. The earliest stirrings of germ theory relate to malaria, attributed to the prolific Roman scholar and soldier Marcus Varro (116–27 BCE) in *Res Rusticae* (Country matters), a book on farming. He wrote, 'Precautions must also be taken in the neighbourhood of swamps [...] because certain minute creatures grow there which cannot be seen by the eyes, which float in the air and enter the body through the mouth and nose and there cause serious diseases.'

The malarial bite is a link in a chain that loops *Plasmodium* microbes, the gut walls of female mosquitos and the human bloodstream. The loop doesn't begin with *Plasmodium*, in so far as *Plasmodium* cannot begin by itself; it is part of a large group of endoparasitic organisms which have long since lost the ability to live freely outside of their hosts. In the lead up to the bite, the *Plasmodium* zygote embeds itself in the mosquito's gut wall. There it develops into a cyst and produces infectious cells, which make their way to the salivary glands of the mosquito.

When the mosquito locates exposed human skin, her bite flushes the infectious cells into the human bloodstream. These cells then embed themselves inside red blood cells, feeding on iron from the haemoglobin. In a more developed form they undergo a rapid mitosis that ruptures the cell and releases them into the blood, where they attack and penetrate more blood cells. This process happens at roughly the same time throughout the human body, manifesting as the periodic attacks of malarial fever.

Eventually, the process produces male and female gametes, which must be taken back to the mosquito's gut to be fertilised. In the case of malaria, as with many zoonotic diseases, 'the bite' has become something of a synecdoche for the whole cycle. Yet, when an uninfected mosquito bites an infected human, their gut becomes *Plasmodium*'s new incubator – *we* pass the parasite to *them*. It's a cycle of transmission after all, not a chain. Like Dickinson's poetics, this instance of my body working against itself complicates the usual narrative. If the bite is the moment where I am a victim, then my passing the parasite back to the biter is when I am, in spite of myself, actively being my own monster.

A lot has been written about infectious diseases and the figure of the monster, tying the latter's appearance in popular culture to the vicissitudes of disease and public health crises. In this lineage, werewolves manifest the infectious terror of rabies, and vampires embody the twofold fear of the transmogrifying bite *and* the partial contamination of the vaccine. It has recently been suggested that the twenty first century's obsession with the zombie is a reflection of emerging 'skin-eating' infectious diseases like Zika and Ebola. What these monstrous creatures have in common is a capacity to transform the victim into their unhuman likeness; as Eula Biss puts it, 'What makes Dracula particularly terrifying [...] is that he is a monster whose monstrosity is contagious.'

The idea of contagion is perhaps one way to acknowledge our own latent monstrosities. But it also keeps us innocent, and passive. Maybe that's why I feel the need to counter it with something more implicative – a cellular version of what the French critic Hélène Cixous calls 'love of the wolf'. Her essay of the same title is an enzymatic treatise on love's metabolism: who eats and who is eaten, who goes hungry and who gets fed. Love of the wolf is the intertwining of love and fear, the thrill of nearly being eaten, the admission that 'some of us really like what scares us':

> The wolf says to the child: I'm going to eat you up. Nothing tickles the child more. That's the mystery: why does the idea that you're going to eat me up fill me with such pleasure and such terror? It's to get this pleasure that you need the wolf. The wolf is the truth of love, its cruelty, its fangs, its claws, our aptitude for ferocity. Love is when you suddenly wake up as a cannibal, and not just any old cannibal, or else wake up destined for devourment.

Love is mixed in with death, the eater rises out of the eaten. Love of the wolf is an interpretative key to the appetites of fairy tales, in which someone or other always gets gobbled up, but it also functions as an ur-fairy tale about multicellular origins, when sex first arose from the failed cannibalism of single-celled organisms. 'When there is a danger from the outside, you bolt,' Cixous writes, 'but when the danger comes from inside, how can you bolt? The danger from the inside is that complicated thing, the love of the wolf, the complicity that attaches us to that which threatens us.'

In a complex cycle of transmission, the bite appears as a sign from the outside: an injury, an attack. The puncture of the skin would appear to reinforce the dichotomy of inside and outside, self and other – I could hardly have bitten myself. But in a way, I have. Becoming a part of each other is monstrous, is 'danger from the inside.' In Cixous's understanding this complicity – or, I prefer, vulnerability – is also what enables us to love. As complex multicellular organisms arising from dozens of monstrous symbiotic mergers, we are susceptible to interspecies love of the wolf; we are, as Cixous writes, 'born-eaten'.

I find a correlate to this in parasitic immunology, but to get there, I have to go back a bit. There was a moment in the heady bacteriomania of the late nineteenth century when every ailment was believed to be of bacterial origin. Viruses had not yet been discovered – their tiny particles continued to slip through the fine sieves devised by scientists – but no matter: the idea of having identified bacteria as the disease-causing kingdom was reassuring. Every pathogen could then be cultured and controlled in the lab, and a vaccine developed. It was something of a disappointment, then, when parasitic protists were also discovered to transmit diseases. The protists are more complex organisms than bacteria – larger and more diverse in morphology, motility and behaviour. They are also much harder to vaccinate against.

An epidemiologist once described to me the process of developing a vaccine with reference to Lego bricks: a (stable) virus is a Lego block, a bacterium is a Lego house, and a parasite is a Lego city. The more complex the organism, the more difficult it is to develop a vaccine. That is why so many diseases caused by protists – like malaria – can become embedded in the body as a chronic infection. Once established, protist parasites have a number of defence mechanisms to avoid being expelled. These include attempting to avoid host detection by living in an intracellular habitat (such as the malaria-causing *Plasmodium* inside red blood cells) or suppressing their host's immune system by affecting the antibody and cellular responses to any foreign antigen.

Some parasites display a third mechanism to evade detection, an 'antigenic mimicry' in which the parasite avoids immune detection by coating itself in the host's macromolecules. The 'masking' of 'foreign' agents makes it all too easy to use military metaphors of decoy and camouflage, to cast the protists as undercover wartime infiltrators that deceive the innocent immune system. It's another worn-down trope – the insider exposed as outsider, a wolf in sheep's clothing. Other sources I consult about antigenic mimicry are more ambivalent, describing it as a 'structural similarity' between parasite and host, or a 'sharing of antigenic sites'. Since the parasite and its host are both composed of cells with nuclei, it follows that they have similar biochemical pathways and resource requirements.

In a parasitic dynamic, this 'structural similarity' forces the parasite to compete with its host for energy and nutrients. I resent the Darwinian inevitability of this, easily co-opted as a fable of natural resource scarcity in which a *Hunger Games*-style fight for survival is normalised. In my research I often came up against material like this, biological processes that appeared so haunted by their interpretations as to seem inseparable from them. These interpretations were usually at odds with my politics; I cannot subscribe to resource scarcity as a 'state of nature' when it is something both created and weaponized by capitalism as a means of control. So I took my deviations where I could, which was often to pick at the language that was presented to me as neutral, objective and precise.

Some threads came loose easier than others. In this instance, the fray of a contradiction appeared. If hosts and parasites are so structurally similar as to have the same biochemical pathways and resource requirements, then how much sense did it make to describe parasites as enacting 'mimicry' and 'masking' in the sharing of antigenic sites? Walter Benjamin describes the mimetic faculty as the 'powerful compulsion [...] to become and behave like something else'. Mimicry supposes difference, rather than similarity, and so in this context struck me as a misnomer. I can't deny that parasites coat themselves in host tissue, but I think the immunological descriptions dispense with their admission of host-parasite similarity too quickly. The becoming-other of antigenic mimicry sounds less like parasites imitating me, and more like them *becoming* me, troubling the grouping of which 'I' am composed. Is mimesis simply a copy, an imitation, or is it – as the anthropologist Michael Taussig suggests – a sympathetic magic, transforming the world as we know it?

As Deleuze and Guattari remind me, becoming is a two-way street. If my parasites are becoming me, I am also becoming them. In *Powers of Horror*, Kristeva draws on Freud's famous case study of Little Hans, a five-year-old boy crippled by his phobia of being bitten by a horse. Is Little Hans scared of being bitten, Kristeva asks, or is he afraid of biting? The victimhood of phobia is problematised through exploring the inverted projections of one's own destructive or monstrous needs and desires. 'Does not fear hide an aggression,' Kristeva asks, 'a violence that returns to its source, its sign having been inverted? What was there in the beginning: want, deprivation, original fear, or the violence of rejection, aggressivity, the deadly death drive?'

In Kristeva's interpretation, as in work by Melanie Klein, fears and phobias derive from early fantasies of escaping fear and lack through incorporating the other, namely, the parental breast. Reading both Klein and Kristeva, I am struck by the frequency with which biting and devouring are mentioned. Klein's ten-year-old patient Richard also suffers from a number of acute phobias, including a terror of other children. During their sessions, Klein observes Richard's tendency to pick up objects in the room and bite them: 'He admitted that he was aware of his tendency to bite; when he felt angry he often wanted to bite, and made biting movements with his jaws, particularly when making faces. As a little boy he had bitten his nurse. When he had a fight with his dog, he bit it if it bit him.'

Klein doesn't withhold her interpretations from Richard; she interprets that 'he wanted to burrow himself with his teeth' because as a baby 'he might have wished to burrow himself into Mummy's breast and devour it'. Like Little Hans and the horse's bite, Klein postulates that Richard's fear of his own destructive appetites, transformed through a process of inversion and external projection, is the source of his phobia. The acknowledgement of these 'monstrous' needs leads to a more integrated sense of self. As a result, the tyranny of the phobia over the psyche will become diminished.

Love of the wolf is a monstrous desire; only a monster could want it. I've since wondered if Lee's cousin suffered an infection as a result of her microbial intimacy with Mutt. But bestiality doesn't have to come into it for things to get monstrous, or rather, everything is already bestial at a cellular level, like Donne's 'three lives in one flea' or the *Toxoplasma gondii* – another protozoan parasite – it's very likely that I pass back and forth with my cat. Now latent, as it is estimated to be in between 30 and 50 per cent of the global population, this pathogen could bloom into a full-blown infection at any point. My little beast could never! Love of the wolf rewilds the myth of harmless domestication, reminding me that monstrosity is always close at hand. Even if my good, Oedipal animal never bites me, she can never be split off from her microbial profile that is irrevocably interwoven with my own. Infection and affection grow along this intimacy that both binds us, and contains the potential to undo us. Perhaps what I fear in the bite is not the infectious wound or the stigmata of fleshy commonality; perhaps I fear the bite that stymies biological flow, the bite that puts an end to my own biting, the bite through which a part of me is bitten clean off.

I Lose My Head

We're beheaded by the nick of time.

– Annie Dillard

I sit down to write about love and no words come out because love makes me lose my head, and my head – I've been told – is where the words are. The fridge is whirring.

I hear children's shrieks from the primary school down the street, and my eye lingers on the new growth on the silver birch outside my window, whose head was severed before I moved to this flat a few months ago. I am used to seeing tree limbs removed close to the trunk, smooth pollarded coins turned to cicatrices – watchful wounds. But here the entire upper part has been sliced off in a clean horizontal line, as if the tree has been decapitated.

Or perhaps it's only possible to lose your head if, morphologically speaking, you have one. The birch's equivalent to a head is dispersed somewhere underground, in a mycorrhizal spooling with the foundations of my building. (Now the birch is rustling its leaves, as if rolling its chlorophyllic eyes.) I associate the word *deadhead* with my grandmother's long skirts patrolling her garden at dusk, secateurs in one hand, excised flowerheads in the other. She was merciless, her blades descending at the slightest sign of wilt. It was a garden filled with hydras, an immanent populace of heads that thrived and multiplied under her rule.

Unsurprisingly, it took a long time for me to prune anything myself. It's a karmic instinct: don't dish it out if you can't take it. Like anyone who couldn't survive their head being cut off, I recoil when reading about the judgement of King Solomon. In this old testament story, two women who live in the same house claim to be the mother of a baby. Solomon listens to each of them before taking up his sword. He proposes splitting the baby in two so that each woman may have half each. Solomon's sly appeal to corporeal individuality – literally, that which is in-divisible – exposes the false mother, who accepts his proposal of division, and identifies the real mother, who offers the baby to her rival in order to save its life.

Unlike me and the contested baby, plants are dividuals, able to survive being divided. The cognitive leap from animal individuality and plant dividuality may be one of the biggest challenges in horticulture – to cleave division from a sense of violence and ally it to care and regeneration. Perhaps it was refusal of this violence (and not just a belief in transspecies reincarnation) that impelled the presocratic philosopher Empedocles to cry, 'Wretches, utter wretches, keep your hands from beans!'. Building on the work of Pythagoras who had taught that eating animals was self-mutilative, Empedocles extended kinship to plants, suggesting that plants with torn leaves experience 'nudity', and a kind of suffering.

But as plant philosopher Michael Marder writes, empathy with plants is a difficult business, rife with anthropic projections. I imagine a plant fears its beheading just as I would in its place, but this, Marder might remind me, is empathy based on a principle of similarity, rather than difference.

I would like to know how successful Empedocles was at refusing the heterotrophic diet – perhaps he found a way to live directly off the sun and the moon, or was able to locate nutritious loopholes within the plant and animal kingdoms? Even if I abstain from eating members of the animal kingdom, my appetite spills over into neighbouring kingdoms of plants and fungi. I am, in a very real sense, doomed to consume. But as any ecologist or presocratic will tell you, matter is always on the move, and consumption never has the last word. We live and breathe, after all, in what Tyler Volk calls a 'wasteworld' of other lifeforms' metabolic by-products. In her short memoir *Love's Work*, philosopher Gillian Rose has a way of putting this imbrication with waste. Shit may be shit, but it's also, as Empedocles might have fantasised, 'the hourly transfiguration of our lovely eating of the sun'.

❂

Losing your head in love is considered to be a rejection of the logical and rational, a state incommensurate with the reality of wider circumstances. The lover's auto-decapitation is part involuntary, part voluntary; part wound and part wounding. In a well-known story from the *Decameron* love is a guillotine, literally and figuratively. Three merchant brothers discover that their younger sister, Isabella, is in love with one of their colleagues, Lorenzo. Far from a discreet courtship, Isabella and Lorenzo enact a frenzy of passionate encounters, one of which is witnessed by Isabella's brother. Like the goods in their warehouse, their younger sister is an object to be traded at a later date; they need her intact. And so they take Lorenzo out to the countryside and murder him.

But Lorenzo, as the dead are wont to, appears to Isabella in a dream. He tells her of his fate and where his body can be found. Grieving but lucid, Isabella exhumes his body, but finds she is unable to carry it back unnoticed. Conducted by love's pragmatic brutality she cuts off Lorenzo's head and smuggles it home, where she places it in a large pot along with soil and a young basil plant. Fed by nutrients from her decomposing lover and the tears she sheds over the pot, the basil plant thrives and blooms. This symbolically romantic gesture is also microbially expedient, since teardrops contain lysozyme – an antimicrobial protein whose targeting of bacterial cell walls works similarly to penicillin. Isabella's tears may not bring Lorenzo back from the dead, but they become a way to tend to his decay.

The flourishing of grief and the basil plant does not go unnoticed by Isabella's brothers, who see 'she [is] losing her looks and her eyes ha[ve] become sunken in her cheeks.' They confiscate the pot and discover its incriminating contents, which they bury for a second time. Isabella, Boccaccio tells us, 'never stopped crying and begging for her earthenware pot' and dies weeping. In discovering and losing her lover's head she also loses her own, never to be recovered.

Perhaps it's not so much that love makes me lose my head as that I temporarily forfeit one modality in favour of another; the sovereignty of the head becomes dispersed throughout the body as the brain releases tides of phenethylamine, dopamine, oxytocin and cortisol. When cortisol enters the bloodstream it causes the blood vessels around the gut to constrict; meanwhile my gut flora may harbour their own long-term amorous wishes – studies suggests that the microbiomes of lovers who cohabit are more diverse than those living alone. In this sense, being in love might be the closest I come to a plant-like dividuality; I lose my head but I do not die. I become quasi-autotrophic, converting love into energy, for the short term at least. Our gut flora usher us into union, and we in turn – like the basil blooming from Lorenzo's head – verge on the vegetal.

○

The head has a lot to answer for, perhaps too much. Sensing danger, the heart orders decapitation of the organ that would seek to curb its passions. *Off with her head!* cried the Queen of Hearts. What was the threat posed by Alice's head? Perhaps that it might overrule the heart, the fickle domain of the Queen? Or perhaps the Queen of Hearts has merely been underappreciated as a horticulturalist; if the head is figured as the Sovereign, king of the body, the Queen is simply meting out to Alice the tough love of deadheading.

In the blazon, a late medieval poetic form, the beloved's body is chopped into parts in order to extol its whole. The blazon rooted in England and grew through Elizabethan literature; in a poem by Thomas Campion, the speaker's beloved 'has a garden in her face / Where roses and white lilies grow'. With a blazon – also a heraldic shield or coat of arms – the poet is able to dismember his beloved's body, shielded from the reality of her blood and guts and inevitable decay by virtue of his poetic devices. In any case, it seems the beloved often survived this mutilation too – Italo Calvino relates a sonnet by Cavalcanti, in which 'the body is dismembered by the sufferings of love, but goes on walking about like an automaton'. Plant-like, the beloved's body thrives through division.

Consulting the blazon's logic, some lovers have attempted to cure themselves of love by removing the body parts in which they believe its infection is located; as if, like bindweed, the root may be unearthed and cauterized. This view of carnal love as a localised infection was one shared by the Desert Fathers, who sought to geographically isolate themselves from sources of temptation by retreating into the wilderness. Others, on realising that they still carried the source of temptation with them, resorted to self-mutilation or what one scholar has called 'radical corporeal asceticism'.

Origen is said to have castrated himself (or paid a physician to castrate him) following his interpretation of Jesus's words in Matthew 19:12: 'there are eunuchs who have made themselves eunuch for the sake of the kingdom of heaven'. The contagion metaphor – and its corollaries of quarantine and excision – had been a crucial underpinning of Christian theology since the fourth century, when the early church father St Jerome uttered a string of contagion metaphors: 'the heretic should be cut off like a piece of dead flesh, expelled from the fold like a sheep with scabies, lest the entire household, body or flock should burn or rot.' Carnal love's mascots in the body, 'the parts of shame' were understood – and treated – as infectious corporeal heretics.

In medieval burial practices, the Sovereign had two bodies. The body politic was a public, spiritual metaphor of sovereignty and dominance, while the natural body was the ruler's corporeal, individual self. (Hence, 'The king is dead. Long live the king.') Upon burial, the natural royal body was often dissected and divided, its organs bound for separate resting places. Sovereignty is a state of dividuality, a blazon made flesh, a body able to survive its own division by splitting into the natural body and the body politic. This raises the question: how do you kill a king if his body politic lives on? If the dismemberment of his body becomes a blazon, whose parts extol his whole?

On my desk is an image I have long been entranced by: the headless man that Bataille named Acéphale (from the Greek for 'headless') gazes up at me from the sunken eye-sockets crowning his groin. This figure gave its name to the secret society and journal formed by Bataille after severing with the main Surrealist group and their domineering leader, André Breton, in 1936. Acéphale was more incendiary in its iteration of Surrealism, organised around a political and pseudo-religious image of decapitation as revolutionary gesture: regicide and deicide rolled into one. Acéphale is the twofold decapitation of the *natural body* and the *body politic*; the acephalous individual is one freed from the external sovereignty of God and king, as well as from the prison of his own head that signifies their internalisation. The text on the back of the journal's first issue is a call to arms, ending in the declaration that 'man will escape his head like a condemned man escaping from prison.' A more contemporary way of putting Acéphale's entreaty might be to *kill the cop in your head*.

I glance back at the birch outside my window and see another headless man. It's the point in autumn that seems both early and late; in either case the leaves are still thick and not yet fallen. All I can see of the tree's crown is three paper-white limbs poking above the canopy, like mandrake roots. The headless birch seems to animate a poem I love by Lorine Niedecker:

> Spring
> stood there
> all body
>
> Head
> blown off
> (war)
>
> showed up
> downstream
>
> October
> is the head
> of spring
>
> Birch, sumac
> before
> the blast

Niedecker's rendering of an Acéphale-like figure appears to represent earth's seasonality through the distortion of human brutality, '(war)'. But if autumn is the head of the body, the poem also seems to suggest that it will naturally fall – or auto-decapitate – as autumn makes way for spring. Is it brutality Niedecker traces in 'the blast' and 'blown off', or simply a changing of the seasons? The poem's tension between finality and cyclicality echoes that of individuals and dividuals; spring can survive its head being blown off, a man gone to war cannot.

Bataille's Acéphale is similarly – symbolically – terranean in nature. An earlier text, *The Solar Anus* (1931) charts the violent movements of love across matter, a dark *ars erotica* of the earth which erupts from volcanic orifices and 'jerks off in a frenzy'. These erotic convulsions are figured as coterminous with the performative vitalism of language: 'when I scream I AM THE SUN an integral erection results, because the verb to be is the vehicle of amorous frenzy.' In both texts Bataille draws on his study of medieval allegory which symbolically allies the figure of the Sovereign with both the head and the sun. So the headless figure of Acéphale is the autumnal twin of *The Solar Anus*; the Sovereign is decapitated, and in place of the head the sun is coupled (copulates) with its opposite, the anus. The violence of the beheading is crucial, ensuring that the allegory's hierarchy is not just inverted, as the more arterial iterations of Surrealism would have it, but decapitated – blown off.

In the shadows of these figures of headless men and debased suns lurks the post-revolutionary context of Bataille's writing: the Reign of Terror whose blades dispatched the heads of a lineage of Sun Kings. The eighteenth-century nobility was bloating at the top of the food chain, and the peasantry had had enough. Jean-Jacques Rousseau may have fabricated Marie Antoinette's dictum to the peasants starved of bread – *Let them eat cake* – but he was following the right trail of crumbs. *Parasite*, after all, had been a common term since the Middle French of the sixteenth century, which, via Latin, drew on the Greek *parasitos*, meaning 'beside the grain'. It was a term used in social relations long before it was imported to ecology, referring to 'one who eats at the table of another'; 'one who lives at another's expense'. The first recorded use of parasite in a scientific context – denoting a parasitic relationship between species – dates to the 1640s. But the effects of parasitism are well known by the parasitised long before they have a name for it; those at the bottom of the food chain don't need to know the vocabulary and its shifting contexts to feel its weight.

This slippage between social and ecological contexts, human and nonhuman, would appear to be confirmed by my bacteriologist friend Hans Zinsser. He writes that parasitism is not just a human state of affairs, but applies to all life on earth with its 'endless chain of parasitism'. It's a state, he writes, that 'would soon lead to the complete annihilation of all living beings unless the incorruptible workers of the vegetable kingdom constantly renewed the supply of suitable nitrogen and carbon compounds which other living things can filch. [...] In the last analysis, man may be defined as a parasite on a vegetable.'

I find a similar position in the work of philosopher Michel Serres. For Serres, parasitism is unavoidable, a set of natural relations that characterise humanity's inherent behaviour as a species. In his book *The Parasite*, Serres draws on an additional meaning of the word in French which is absent in translation; parasite also denotes *noise* or *interruption*. Thus human relations form 'a parasitic chain which interrupts or parasites other kinds of relations (that is, those of other animals, or the natural world itself)'; to be human is to be heterotrophic, an interrupter of nonhuman resources to survive.

Serres's and Zinsser's admission of the ubiquity of parasitic relationships rings true, although there's something that bothers me about Serres's book. I think it's connected to that tension in Niedecker's poem about fate, about the ones who get sent to die in war versus the ones who make the call and get to live. It's as if to acknowledge the universality of parasitic relations would be to somehow let wealthy social parasites off the hook, to normalise hierarchies and exploitation. I give myself whiplash reading like this, oscillating between what Eve Kosofsky Sedgwick would call 'reparative' and 'paranoid' readings of Serres's book, pulling back from the text and then tentatively resuming, as if trying to catch a latent interpretation in its moment of transmission.

Giving Serres the benefit of my doubt, I wonder if *The Parasite* is less a justification of inequality through an attempt to naturalise it – like Social Darwinism – and more of an existential reckoning with a fundamental interdependency that muddies the distinction between 'guest' and 'host'; in French the word *hôte* refers to both. Parasites have parasites which in turn have parasites. Serres writes that this intersubjectivity is the 'atomic form' of our relations, and rallies a cry reminiscent of Bataille's allegories: 'Let us try to face it head-on, like death, like the sun. We are all attacked, together.' This 'atomic form' of relations brings to mind Thomas Browne's *quincunx*, the net or mesh of which the seventeenth-century author believed life and creation was composed. Drawing on classical and gnostic ideas of five as a sacred number, the quincunx evokes life as a net or web, rather than the traditional tree of Darwinian evolution. Instead of filial and hereditary branches, the quincunx is a levelling of relations that spool out in all directions.

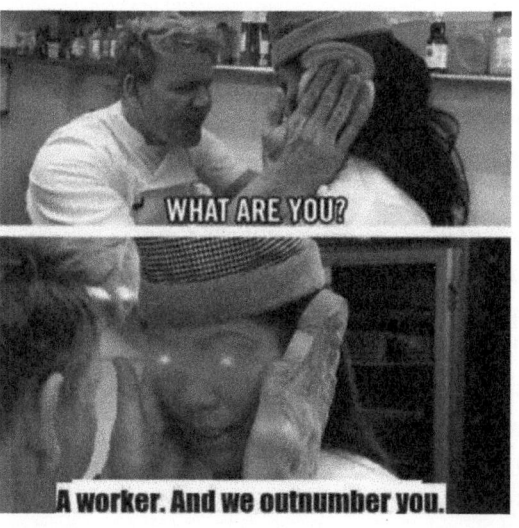

Parasitism, after all, is an inversion. Hierarchies suggest a concentration of power at the top of the chain, but the most powerful organism in the world can be undone by a parasite an infinitesimal fraction of its size and metabolic prowess. The rich may be parasites on the poor, but the poor are the multitude, embodying the potential of ecological parasites to undermine organisms far more powerful than themselves. Ecological parasites become unlikely comrades when they flout social hierarchies, proving Hamlet's point that 'a king may go a progress through the guts of a beggar':

> Your worm is your only emperor for diet: we fat all creatures else to fat us, and we fat ourselves for maggots: your fat king and your lean beggar is but variable service, two dishes, but to one table: that's the end.

In 2013 *The Lancet* published a report further proving this point. Eggs of the large roundworm *Ascaris lumbricoides* had been found in the pelvic area of Richard III, whose remains had been excavated from a carpark in Leicester the previous year. *Ascaris lumbricoides* is one of the largest roundworm species, with males growing 15–30 cm long in the small intestine and living for up to a year. Symptoms of living with *Ascaris lumbricoides* vary from mild abdominal discomfort and indigestion to painful duodenal ulcers. Some specimens have been known to 'wander' from the small intestine to other organs, causing further complications such as appendicitis, or blockages of the bile duct which bring about jaundice-like symptoms.

Intestinal parasites – roundworms, tapeworms and liver flukes – were near ubiquitous in late-medieval and early-modern Europe, leading the seventeenth-century physician William Harvey to remark of the English: 'Worms in all the guts; nothing so common as worms'. John Donne's final sermon dwelled on the horror of 'vermiculation'

– being eaten by worms in the grave, evoking the endless churn of matter: 'In the grave the worms do not kill us; we breed, and feed, and then kill those worms which we ourselves produced.'

Gut worms were common, but in the body of a Sovereign they were also an unruly commons, a shared condition that paid no heed to social status and hierarchy. When the life cycle and transmission route of an infectious agent – bacteria, virus or parasite – is known, it follows that those with the most wealth and resources can afford the strongest line of defence available. But parasites are endlessly inventive; the cooks preparing Richard III's food weren't to know that the eggs of *Ascaris lumbricoides* are easily transmitted via faecal contamination, or that said eggs can survive – dormant – outside of their host's body for up to five years.

Richard III wasn't beheaded, but perhaps his gut played a leading role in his 'winter of discontent'. The gut–brain axis has since established the interrelation of gut health and mood; over 90 per cent of the body's serotonin – responsible for, amongst others, feelings of happiness and wellbeing – is located in cells of the gastrointestinal tract. Edith Sitwell could have been talking about Richard III's riddled microbiome when she called Shakespeare's play about the monarch a 'Ritual of the Falling of the Sun'. The royal microbiome was a decorticated *natural body* whose *body politic* strove to overcompensate, 'determinèd to prove a villain'.

❂

Acéphale seems to know what plants know, that sometimes you have to lose your head to move forward. I understand this, but deadheading still unnerves me – what if the head never grows back, and the torment of losing it is for nothing? These questions plagued me while watching Shane Carruth's experimental sci-fi film *Upstream Color* (2013), which follows the three-stage life cycle of a fictional parasite. It begins with a character known only as the Thief, who harvests the parasite in its larval stage from the soil of plants and secretes the worms inside pill capsules. At a nightclub we watch him taser and feed the pill to a young woman named Kris. The effect of the parasite is immediate: Kris is reduced to a naïve, trance-like state. Under the Thief's influence Kris drives him to her home and waits listlessly in the doorway of the kitchen while he gives instructions: she must not eat because all the food has been poisoned; she must stay up all night transcribing pages of Thoreau's *Walden* onto note paper, which must then be folded up and made into paper chains. Upon the completion of each paper chain, she is permitted a small amount of water. Kris performs this ritual obediently throughout the night while the Thief sleeps.

The Thief doesn't just tell her what to do but also what to feel; he tells her that all she wants is water, and that she has never tasted anything so refreshing in her life. The Thief's words are hypnotic; Kris internalises them as her own. He tells her she cannot look at him directly because he was born with a disfigurement which means his head is made from the same material as the sun. Kris looks down at her paper chains, but her gaze is unfocussed and absent, it is perhaps no longer Kris that sees them. The following day the Thief pretends to receive a call from Kris's mother – he tells her she has been taken by several men who are demanding money to set her free. Kris is not so absent as to be undisturbed by the news. 'Oh no,' she intones distantly, and asks the Thief if he has any money. 'I don't,' he says. 'Do you?'

What follows makes for painful viewing, as a half-starved Kris deliriously empties her savings, signing cheque after cheque to the Thief, and unearths a stash of precious coins from beneath the house. In one scene, the Thief lays Kris's outfit out on the bed before they drive to the bank where she will sign away the deeds to her home. In the car they rehearse the lines she will parrot to the mortgage advisor; she laughs with the Thief and appears to be happy, enjoying the game. Days later, when Kris comes to and cannot remember anything to do with the Thief, the bank provides her with copies of the documents with her signatures, and CCTV footage of her inside the building. Kris stares at the grainy images incredulously. It is her but it wasn't her; if she doesn't understand what happened, how can she explain it to anyone else?

When the Thief abandons Kris, things get worse before they get better. She gorges on food and falls asleep; when she wakes up she sees worms, several inches long, coursing beneath the surface of her skin. Her frenzied eating also fed her parasite, which has now matured in her body. After unsuccessful attempts at gouging out the worms with a kitchen knife, Kris tracks down the Sampler, a reclusive pig farmer who is familiar with the parasite. He hooks Kris up to his homemade machinery and, through a process of painful extraction, transfers her parasite into the body of a pig. When Kris wakes up on the side of the road she once again has no memory: she doesn't understand the knife wounds all over her body, and on returning home is confused by the mess of food and blood in the kitchen. She returns to work only to find she has been fired due to her unexplained absence, and a declined card at the supermarket checkout is the first inkling she has of her financial state.

All this takes place in the first half-hour of the film. *Upstream Color* is slow-moving and narratively elusive; intermittent shots of pigs on the Sampler's farm slowly unfurl the premise that Kris is intimately connected to the pig who carries her parasites, and that the other pigs on the farm are similarly bonded with those whose parasites the Sampler has extracted. When he drowns the piglets in a river, a noxious liquid leaks from their decomposing bodies and grows into blue orchids that appear along the river. The plants are then picked and sold, enabling the Thief to harvest the soil for fresh larvae and perpetuate the parasite's life cycle.

The fictional parasite in *Upstream Color* is nameless, and its three-stage cycle between human, animal and plant bears little resemblance to any known parasites. Yet the erosion of Kris's identity and agency is a compelling depiction of parasitic possession, of losing your head. The particular power of the parasite is that it is not always obvious you are under its sway; like the signatures on Kris's documents, you can still appear to the world, more or less, as yourself.

Around the time I first watched this film, friends were sending me links to a clip that had recently gone viral. *These Worms Turn Snails Into Disco Zombies* is a compilation set to a minimal synthy soundtrack. It shows various snails which all seem to have grown brightly coloured Haribo sweets inside their eyestalks, which pulse up and down in time with the music. *No, this snail isn't ready to rave* ... says the voiceover. The Haribo sweets, it transpires, are the larvae of a parasite – *Leucochloridium paradoxum* – which have gorged on the snail's insides and are now using its eyestalks to masquerade as caterpillars. Like many endoparasites, *L. paradoxum* needs a warm digestive system in which to mature, and this parasite's choice guts are those of crows, jays, sparrows and finches. Inside the birds' digestive systems, the mature parasites lay eggs which pass out through droppings and are subsequently devoured by their intermediate hosts – land snails in the damp forests of Europe and North America.

The snails are much more unfortunate than the birds. Inside their digestive glands, the parasite eggs hatch into larvae, which grow and push their way into the snail's eyestalks, overriding the brain and colonising up to one third of the snail's body. The snail is now a zombie, whose pulsating eyestalks mimic a bird's eye view of a wriggling caterpillar and guide the snail into exposed areas where it can be easily spotted. There is no defence for the snail when a bird catches sight of it; the unsuspecting bird pecks off the eyestalks, delivering the larvae to its digestive system, where the cycle can begin again.

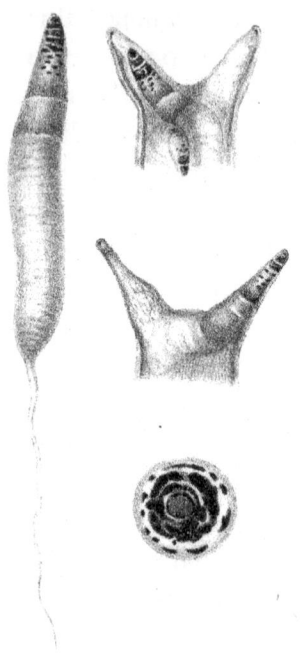

The snail survives the loss of its eyestalks, which eventually regrow. But the unfortunate snail is infected with the *Leucochloridium* parasite for life – doomed to regrow its eyestalks which will be pecked out by birds and regrown once more. The snail's fate is unequivocally Promethean, recalling the overreaching mortal whose punishment was to have his liver pecked out by an eagle each day and regrown each night, in a never-ending cycle of agony.

The measure of a parasite's success is its ability to keep both itself and its host alive – if the latter dies, it takes its parasites with it. Like the unnamed worms in Kris's body, *Leucochloridium* species are successful parasites, hollowing out their hosts' bodies and using those resources to their own benefit. Where Prometheus differs from *L. paradoxum* is his relation to narrative. Though both agonies are cyclical, Prometheus' fate is understood as a grisly repercussion for the trickster who marred the divine order by stealing fire from the gods. His splayed liver enacts a kind of haruspicy, divining cause and effect, action and consequence. The narrative is inexhaustible, able to be pecked out each day and regrown overnight. But the snail infected with *Leucochloridium* is no trickster; it stays firmly put in its ecological niche. Perhaps this apparent lack of provocation is why parasitic relationships are so difficult to stomach; the mind grasps about for reasons, like eyestalks parasited by the word 'why'.

Medieval mystics had a way of raising themselves up to be pecked out by God; they called it ecstasy. Anticipating the discovery of the human microbiome, or perhaps that the cells of our bodies are a multispecies compost, the mystics seemed to know that being 'human' isn't a state of inviolable or singular corporeal integrity. To fully experience and extend themselves to God they had to be dismantled, disassembled limb from limb. The mystic Marguerite Porete described the 'exalting ravisher who takes me and unites me with the middle of the marrow of divine love in which I am melted.' This melting of the self's effluvia out and into the divine is the alchemy of ecstasy, a word which, etymologically, means standing or placing outside of – being outside of oneself. The nineteenth-century saint John Vianney had an acephalous way of putting it: 'To be a saint one must be outside oneself. One must lose one's head.'

Swapping Vianney's 'saint' for 'human' or even 'organism' maybe brings me a little closer to the fate of the parasited snail. In Annie Dillard's *Holy The Firm*, an unfortunate moth becomes a burning, headless 'hollow saint', a symbol through which unthinkable atrocities flicker in and out of comprehension. Reading by candlelight while camping alone in the Blue Ridge Mountains, Dillard describes how moths would periodically fly into the candle's flame and singe their wings. But one evening, a moth cannot retract herself quickly enough and is engulfed by the flame that burns away her wings, legs and head, leaving only her body, which takes on a terrible utility:

> And then this moth-essence, this spectacular skeleton, began to act as a wick. She kept burning. The wax rose in the moth's body from her soaking abdomen to her thorax to the jagged hole where her head should be, and widened into flame, a saffron-yellow flame that robed her to the ground like any immolating monk. That candle had two wicks, two flames of identical height, side by side. The moth's head was fire. She burned for two hours, until I blew her out.

The flame parasites the moth's body; Dillard parasites the scene for symbiosis with her thoughts, seeing in the acephalous moth 'a flame-faced virgin gone to God'. But Dillard is not in the business of anthropic errors; the moth is no neat vehicle for faith but an iteration of God as 'a brute and a traitor, abandoning us to time, to necessity and the engines of matter unhinged.' To be without these abandonments is to be without faith, a word I interpret in Dillard as the ability to live and love, in spite of atrocity and a world filled with unhinged matter. As her student Maggie Nelson puts it in *Bluets*: 'Whenever I speak of faith, I am not speaking of faith in God. Likewise, when I speak of doubt, I am not talking about doubting God's existence, or the truth of any gospel.'

The first time I saw *Upstream Color* I was not aware of the extent to which I had lost my own head. I didn't inwardly echo, as I do now, the childlike naivete with which Kris signs away everything she has. I was still some months away from leaving a relationship, which I would suddenly – almost comically, like cartoon scales falling from my eyes – come to realise was not what I thought it was. There were no worms in my body except the latent putrefaction of misplaced trust. Some years earlier I had met someone and, within a turbulent month of meeting them, embarked on a new life which included a joint bank account, since we were no longer two people but one: one blazing amalgam of love, defying the lonely cosmos.

Into the account – at their insistence – went my PhD stipend and what was left of my savings, after a debt to their ex had been paid off. Friends were concerned but I wasn't. Love has its own logic, I reassured them; and besides, I had nothing to worry about – this person was far older than me, owned their own home, received an income from their family's estate. In time it became clear that this income was dwindling, and my stipend was what was keeping us afloat. When I left I believed I would cut my losses, make a clean break. But it transpired that legally I had incurred a kind of debt, since I had lived in their property, thereby benefitting from its capital, which was required to be repaid in full. For the next eighteen months this cost exceeded my rent, and the stress of meeting it became my own strange centripetal hell, spinning inside the larger hell of lockdown.

In hindsight I can see they were not a very good parasite; it would have gone better for them to attach themselves to someone with even a modest amount of familial wealth, rather than the small life-raft I'd stuck together with waitressing and bits of prize money. I was soon sucked dry and then there was nothing left for them to suck; lose lose. But with an even more unsettling amount of hindsight I can see that there was something libidinally thrilling about being destroyed in such a blatant way. It was hard to tell people what was going on simply because it sounded so unbelievable, like I was calling attention to myself as being somehow special for having managed to acquire such a rare and exotic kind of parasite, without having left the suburbs. The intensity of love's ending matched that of its beginning, and there was something in this I must have perversely enjoyed. All boom then all bust, consistently ablaze and then suddenly extinguished, the good breast and the bad, love of the wolf and nothing less.

In Christian mysticism, *kenosis* is a word meaning self-emptying, referring to the descriptions in the New Testament of Christ emptying himself out to make room for God. In a letter to Elisabeth of Schönau, Hildegard of Bingen described a human being as 'a vessel that God has built for himself and filled with his inspiration so that his works are perfected in it.' Kenosis is the body as vessel; a hollowed-out saint, atrophied by love. Outside of its Christian usage, the Greek term can also have negative connotations, used to suggest a bodily depletion and deficiency, or to describe the waning of the moon. The mystic's position is that you cannot be filled with ecstasy and love if you cannot first be emptied out. To be impervious to infection is to cease to be a part of unhinged matter; the snail's kenotic potential is part of being a snail. The nature of having a head contains the possibility of losing it.

We Eat Each Other Up

> *Love is when you suddenly wake up as a cannibal [...]*
> *or else wake up destined for devourment.*
>
> – Hélène Cixous

I lose my head but not my appetite; in love we eat each other up. Part of love's vocabulary is digestive: it consumes and devours, it swallows or else desires to be swallowed up. If love is how we eat each other, we eat in such a way that the meal is inexhaustible. And if one source dries up, as the saying goes, *there's plenty more fish in the sea.*

Simone Weil had a lot to say about love's appetite. She understood humanity as a cannibalistic economy that routinely eats its kin for nourishment. 'We love as cannibals,' she wrote. 'Beloved beings ... provide us with comfort, energy, a stimulant. They have the same effect on us as a good meal after an exhausting day of work. We love them, then, as food. It's an anthropophagic love.'

A few years ago two friends and I proposed a project to an art festival that would consist of a dining experience of Weil's philosophy, incorporating food described in her notebooks: strawberry jam, peaches, egg yolk whipped in sherry, marron glacés and butter rolls. The proposal was rejected on the grounds of its limited audience capacity. We were all used to rejections but on this occasion were admittedly surprised, since contemporary art was at the outset of a mystical turn that would make Simone Weil, among others, its unlikely poster girl. I've since wondered if, from their perspective, our branding was off: foregrounding Weil's writing on food (and the ritualistic pleasure she described in eating it) undermined the popular depiction of her suspected anorexia. Our unrealised proposal, and Weil's cannibalistic economy, lurk in a subgenre I have come to think of as *digestive mysticism*.

Eating is devotional; where would Christianity be without cannibalism after all, without making a meal of God? The Eucharist of my childhood was lacklustre symbolism: Welch's Purple Grape Juice and a rustic-looking loaf from Tesco. I was more smitten with the tenet of transubstantiation in Catholic orthodoxy, in which the bread and wine became Christ's flesh and blood incarnate. What did transubstantiation taste like, I wanted to know – was the blood salty, the flesh seasoned? Did it satisfy, or make you hungry for more? Since I am unconfirmed, I guess I'll never know.

At one point I became obsessed with cataloguing instances of digestive mysticism; medieval scholar Caroline Walker Bynum describes how almost all mystics spoke of tasting God. Many wrote of drinking blood from the breast of Christ, and Mechthild of Magdeburg deemed blood a 'superior food'. Catherine of Siena described 'God as table, Christ as roasted flesh, and the Holy Spirit as waiter and servant.' In 'Love's Seven Names', a poem by the thirteenth-century beguine mystic Hadewijch, the erotic metabolising of God was explicit:

> [...] love's most intimate union
> Is through eating, tasting and seeing interiorly.
> He eats us; we think we eat him,
> And we do eat him, of this we can be certain.

While Hadewijch conceived of humanity turned to food through passionate devotion to God, Hildegard of Bingen was more nutritionally specific. In one of the visions contained within her text *Scivias* or 'know the ways', she correlated people with foodstuffs:

> Because they who are vowed monks are like grain, which is the strong, dry food of humans; so this people of Mine is bitter and harsh to the taste of the world. And the clerics are like fruits, sweet to the taste, and show themselves sweet to people by the usefulness of their office. And the common laypeople are like meat, but meat comes partly from chaste birds; thus those who live in the world according to the flesh have children, but among them are found followers of chastity, such as widows and the continent, who fly to heavenly desires by their appetite for virtue.

Bynum reminds me that, at the time these female mystics were writing, the preparation and serving of food was the woman's sphere. Doctrines forbade them from participating in many aspects of religious life, and so the language of food and consumption – in which they were well-versed – offered symbolic access to the sacred. Eating or its lack was a devotional mode accessible to those marginalised by the society they lived in; the language of the body, of ingesting and digesting, opened up a space of resistant jouissance. For these individuals, not always female but always feminised, what was at stake was often literally *the stake* at which they could be put to death for heresy.

Weil wedded another dynamic to the tradition of devotional consumption: a God who also eats us. Drawing on passages from the Bhagavad Gita, Weil shaped her understanding of a ravenous deity: 'Man eats God and is eaten by God.' This vision of an anthropophagic God makes me wonder what kind of digestive system this deity would possess – perhaps a bovine sequence of stomachs, a metabolism that intermittently regurgitates and ruminates on the cud of humanity? Or perhaps it is God's own offspring, not humanity, that is grass-like. A painting by Friedrich Herlin depicts the medieval conception of Christ as food; a stalk of wheat grows from the wound in his right foot to pierce his left, and from his other foot grows a grapevine. The medieval motif of the mystical winepress expanded this image to depict Christ standing in a winepress, becoming the grapes he crushes into wine for the faithful.

While she believed cannibalism was the force underpinning human relations, Weil also maintained that we are required to overcome it; we must shift our status as *eaters* to becoming *the eaten*. In being eaten by God we are metabolised to become nourishing 'food' for other human beings, thereby breaking the cannibalistic cycle. Weil longed to be devoured by God in this way, but at thirty-four she was also devoured by the bacillus which killed her, *Mycobacterium tuberculosis*, a condition known colloquially as 'consumption'.

When I was in the early stages of writing this book my boyfriend, now ex, asked me if I was drawn to thinking about infection because it is not immediately legible as intentional harm by the microorganisms that cause it. I don't remember how I answered him; I hadn't yet considered my subject in those terms. Our relationship was still young but had already been overburdened, prematurely aged by the parasitical one that preceded it. I had wrongly thought that the financial stresses imposed by my ex were a violence aimed exclusively at me, but I came to realise that they were also intended to harm anyone I got close to. In my boyfriend, that violence found another body vulnerable to its infective power, and we lived the financial stress together, fearing legal letters and fantasised repercussions – bailiffs, debtors' prison – that fed on the financial precarity of our respective upbringings. Compared to the situation we were living through, the violence of microorganisms seemed more indeterminate, diffuse. He was suggesting, I think, that I was dealing with the harm being done to me by focussing on another kind of harm that I could convince myself was unintentional. As if malice – intentional harm – was not something I could bring myself to acknowledge.

I pushed back on my distal awareness of the subtext and told him what I tell myself, which is that I wanted to diverge from the military metaphor, to refuse to rely on a lazy characterisation of pathogenic life as inherently evil, malevolent and vicious. Which wasn't to say the opposite, that pathogens are virtuous beings who can't help but damage others in their own path to survival. Although there was, I admitted, something darkly seductive about this naturalisation of damage. Thinking along these lines led me to illicit interpretations, of Weil's cannibalistic economy extended to the nonhuman, an understanding of the body offered up to nourish other forms of life. Was her own death by bacteria – becoming the *eaten* – the ultimate act of decreation? Was becoming food for other species the definitive gesture towards renouncing our heterotrophic appetites?

But you don't know, my boyfriend said vehemently, *you can't know.* Can't know what? I asked. *Can't know that pathogens aren't malicious. Thinking they're malicious or thinking they're benign or neutral, it's just two parts of the same fallacy.* He was talking about anthropic error, what Saidiya Hartman calls 'the difficulty and slipperiness of empathy', an over-identification with the other which ends up obliterating their specificity and difference. He'd been studying bacterial quorum sensing and knew as much or as little about microorganic life as I did. There was the limit of what we knew, and then there was the subtext. He needed me, it seemed, to keep the possibility of violence in mind, to not look away from damage or to absolve the agents inflicting it, whether they intended to or not. He was reminding me to not underestimate what was being inflicted on me, on him, on us. Much too late I realised he was angry at the infection running through our relationship, and that he was right to be; in the end it proved insurmountable, and consumed us.

If I was the kind of person who thought origins were everything, I could trace the cannibalistic undercurrents of intimacy all the way back to the beginning, when there were no clean lines between eating and fucking. Billions of years ago, before the planet clothed itself in layers of oxygen, the earth was covered in an orgiastic mass of single-celled life. Bacteria writhed in the stink of an anaerobic atmosphere, noxious with the fumes of ammonia, sulphide and hydrocarbons. Like most cells in the human body, these single-celled organisms reproduced through mitosis, what science calls 'asexual' reproduction, since the cell essentially clones itself: it replicates its DNA and splits in two. It takes energy to be asexual, though; the earliest bacteria ate sugars formed by atmospheric phenomena like lightning and volcanoes, and later evolved to draw energy from the sun. But when sugars ran out they starved, and other cells began to look appetising.

According to Lynn Margulis, the earliest stirrings of sex among cellular life evolved from a kind of thwarted cannibalism. One starving single-celled organism tried to engulf another but failed, since it lacked the organs to digest it. Both were kind of alive but no longer quite one or two, resulting in 'a truce called sex'. As Margulis puts it:

> Some cannibals ate and devoured every last cell appendage of their victim brothers. Another might suffer indigestion and spare the nucleus and chromosomes of its intended meal. The two merged cells would form a new single cell with two nuclei and two sets of chromosomes. [...] This was more than aborted cannibalism. [It was] the formal equivalent of fertilisation.

When one organism fused with another in this way, it also took on its genetic material. This doubling effect was cumbersome, inconvenient, a physical strain that is – having never been pregnant – beyond my comprehension. I imagine trying to accommodate another me-sized organism within the current bounds of my body – grim. As Margulis puts it: 'You'd probably yearn for relief from your condition as double-monster.' And so the double-monster located a form of relief by splitting – something it already knew how to do, thanks to mitosis. But this time the cells were not identical; because of the combined DNA of the eater and the eaten, their subsequent splitting created genetic difference.

This merging and splitting is called meiosis – sexual reproduction – and is responsible for the life cycle of animals, plants and fungi. The sexual behaviour of most species no longer displays this cannibalistic heritage, but some seem to wear it on their sleeve, like the well-known devouring lover much beloved by the Surrealists: the praying mantis. For Andre Masson, Max Ernst and others, the mantis was an invocation of erotic violence, epitomised by the female mantis's beheading and consumption of the male during coitus. To the psychoanalytically informed Surrealist mind, this cannibalistic mating ritual made visible the unconscious drives underlying sexual desire. Early naturalists, too, were compelled by this behaviour, studying how different stages of cannibalistic dismemberment prolonged or delayed coitus.

I was surprisingly relieved to read more recent research that interprets these mantis behaviours as stress-related, and that sexual cannibalism is more likely to occur in specimens kept in labs or as pets than in the wild. I was relieved, not because of any aversion to mantis sexual proclivities, but because their appearing to re-enact the cannibalistic origins of sex seemed too easily co-optable, as if they, as a species, offered an unfiltered example of what sex 'really' is. An origin story is a risk, after all: it can illuminate something about the present, but it can also distort it, reducing the present to a point in the past, all the while obscuring the extent to which that past is always subject to interpretation.

I'm thinking here of postcolonial philosopher Édouard Glissant who prioritised relations over origins, since origins and access to them are controlled and authenticated by those with power. In his *Poetics of Relation*, Glissant charts an associative, non-linear approach to writing Antillean memory, culture and experience, reflecting the way in which these have been subject to rupture, discontinuity and expropriation by colonialism. To write – to think – with an unproblematised approach to linear chronology would amount to conspiring with a colonial narrative which conceals the devastation it has wrought, and obscures the break forced between colonised peoples and their pre-colonial pasts.

Glissant is rightly suspicious of appeals to origins, of tracing bloodlines of filiation from a mythic past to the divinely inherited present: it requires the exclusion of impurities. It's similarly impossible to study evolution without being confronted with the violence of interpretation typified by Social Darwinism, eugenics, selfish genes and the survival of the fittest, which propose particular races, adaptations and body-types as the pinnacle of human evolution. This proximity often made me want to reject learning about evolution altogether, as if there was nothing salvageable in it, no way to separate it from the violence it so often gave rise to. Some would say that this was just another anthropic error on my part – that my need to locate possible alignments between my politics and my research was just one side of a coin: socialism on one side, fascism on the other. I was disappointed when I read Lynn Margulis dismissing the 'just-so' stories that rationalise the behaviour of present-day life forms (humans) through thinly veiled comparisons with microorganic ancestors. Disappointed but also confused, since this apparent political neutrality seemed to run against the emancipatory current of her theories.

Symbiogenesis – the theory that Margulis spent a lifetime refining and being attacked for by the mainstream scientific community – is a prime example. It involves the process described earlier, where single-celled organisms began to fuse and merge out of desperation. In some instances, this 'cannibalism' resulted in some cells beginning to act like organs – or organelles – inside other cells. An immediate example is the mitochondria that exist in all animal cells and that are responsible for converting food into usable energy. In the 1960s, Margulis noticed that these organelles bore a resemblance to certain free-living bacteria, and hypothesised that through a series of symbiotic mergers they were incorporated into what we know as animal cells. Now proven, this theory means that, on a cellular level, we are all mutants – hybrids of different life forms that Margulis has described as a kind of 'baroque edifice'.

Many scientists were opposed to this theory because it seemed to threaten the sanctity of Darwinian evolution and its vertical, heritable, linear model of change and adaptation. By comparison, symbiogenesis is horizontal and anarchic, a frenzy of illicit fusions and mergers – energies coming together for mutual benefit. As Martin Brasier puts it, 'Lynn Margulis and Richard Dawkins had learned different lessons from the same book of life. One of them saw networks. The other saw hierarchies.'

The poet Eleni Sikelianos has described symbiogenesis as the 'biological counterpart' to Édouard Glissant's *Poetics of Relation* since it shares Glissant's rejection of access to ancestral origins as the ultimate value. Symbiogenesis as a poetics of relation is good news for anyone with an uneasy relation to the notion of origins. Its horizontality also complicates the verticality of sexual reproduction, much like bdelloid rotifers, another example offered by Margulis. Bdelloid rotifers are freshwater-dwelling microorganisms that look like tiny glass filaments or syringes. Throughout their evolution some rotifer species have remained sexual, while others have simply 'forgotten' two-parent sex and become parthenogenic, sometimes resulting in the obsolescence of males. In studies that compare asexual and sexual rotifer species, the asexual rotifers were shown to be three hundred times more genetically diverse. The bdelloid rotifers therefore complicate essentialist assumptions that sexual reproduction alone accounts for genetic difference.

Lovebug

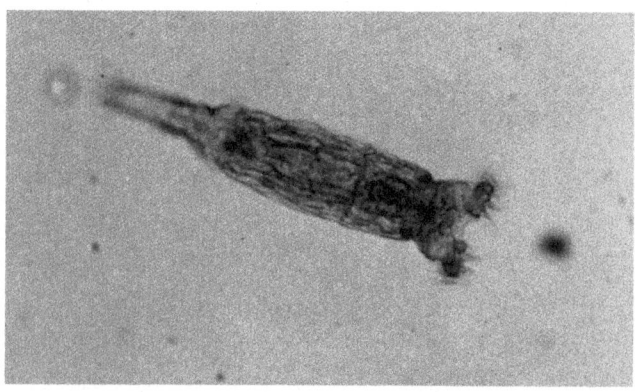

Learning about symbiogenesis and bdelloid rotifers were part of what made me want to keep learning about evolution, rather than look away. If my choice was between conscious anthropic error and the greater anthropic error of pseudo-neutrality, it seemed infinitely better to consciously take politics with me rather than fantasise about leaving them at the door. I needed relations in order to trouble my study of origins. Sex contains the story of thwarted cannibalism but it will also always be in excess of it: its evolution manifests as echoes and traces that loop back on themselves, weaving in and out of the present. Perhaps in our vocabulary of swallowing and eating each other up – in sex and in love – we are voicing ancestral muscle-memory, ventriloquised by the biological unconscious. Sex traces this loss of coherence, a blur of bacterial boundaries, but it will never stop proliferating beyond its perceived origin, becoming something new and irreducible in each encounter.

The fusion of human sex is a fleeting rather than permanent alloy of bodies; we do not stay in position for life. In the psychoanalytic work of Melanie Klein, the notion of permanent sexual fusion has been explored as a source of terror. In one account her patient Richard, an eleven-year-old boy suffering extreme anxiety, admits his fears regarding his parents' sexual intercourse. 'Richard asked if they were stuck together like Siamese twins', Klein reports. He says 'it must be terrible for those twins if they could never get away from each other.' Klein goes on to interpret Richard's anxiety about his parents' fusion in terms of his own burgeoning sexuality – perhaps he fears that by penetrating another's body with his own genitals he will become stuck irreversibly: 'He might never get it out again.'

Like the permanent sexual fusion that Richard so feared in his parents, a pair of adult *Schistosoma* worms are, once mated, joined together *in copula* for life. These worms are parasitic flukes that live in the blood vessels of the bladder, large intestine or small intestine of their human hosts. In copula, the smaller female is threaded through the body of the male, woven into his gynaecophoric canal. Their conjoined bodies form a two-headed ouroboros, a closed circuit of never-ending coitus. The images I find online are beautiful and claustrophobic – it's the kind of image I would have been transfixed by as a child, unable to bleach from my thoughts. As an adult, they remind me of drawings by Louise Bourgeois, particularly her series *À l'Infini* (2008) which comprise an oneiric tangle of fleshy matter: tubes, corpuscles and the gleaming pink hue of histopathology. In their knotted position the *Schistosoma* couple live out their allotted lifespan in the human body – not quite à *l'infini*, but in some cases for up to forty years.

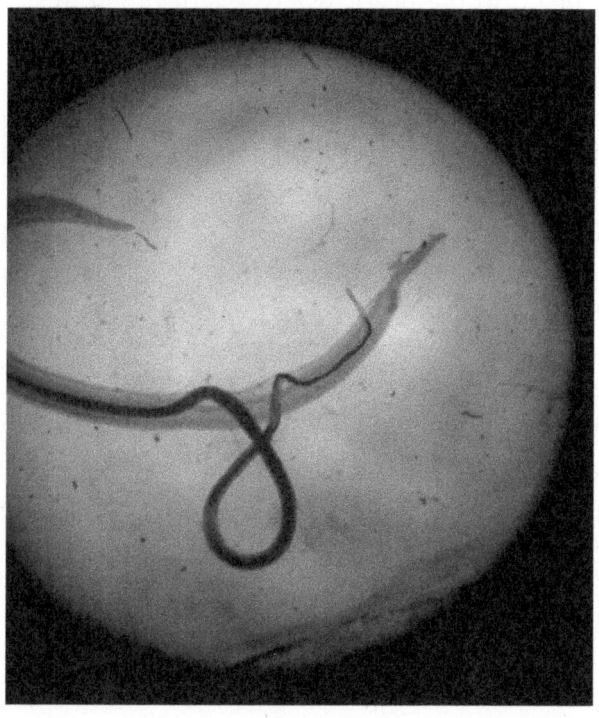

One species makes its love nest inside another; *Schistosoma* worms shack up in the organs of *Homo sapiens* and mate for life. ('This flea is you and I, and this/ Our marriage bed, and marriage temple is'.) How they get inside humans in the first place has been described as a faint lesion, akin to the bite of a flea. It happens in bodies of water infested with schistosome flukes and the snails on which their lifecycle depends. A warm human body enters the water and if they are unlucky a fluke will latch on, burrowing its head into the skin while softening the epidermis with digestive fluids. Once it is through, the fluke rides the venous blood to the heart, where arteries then flush it all through the body. Only flukes taken to the liver survive, where they will take one to three months to sexually mature and mate. The adult schistosomes will then proceed to the bladder or intestine, the final destination of their knotted coupledom.

Now the person – who may not have realised they are a host – will begin to experience symptoms, as the female schistosome releases hard, spiky eggs that claw their way through the wall of the affected organ. The eggs are trying to get into the urine and out of the body, so that the parasitic legacy can continue. This clawing causes agonising pain: but damage to the tissue is caused less by the eggs and more by the inflammation of the body's immune response. In its attempt to halt the eggs' passage, the organ outgrows itself, enclosing the eggs in fibrous tissue. If the eggs manage to pass out through the urine, causing further pain and bleeding, they will hatch and release miracidia: non-parasitic microorganisms. The miracidia will only survive for a few hours, in which time they will need to find their way to freshwater and encounter their intermediate hosts, aquatic snails. Those that do meet a suitable snail burrow into its body and camp out in its intestine while they develop. Eventually they leave the snail and return to wait in the freshwater, until another warm human comes along.

Schistosoma worms derive their name from the Greek for 'divided body'. As a species they appear to have accepted the offer that Hephaestus, god of metallurgy, makes to humans at the end of Aristophanes' speech in Plato's *Symposium*. The short version of this story is that ancient humans were originally spherical with two heads and two pairs of arms and legs. They rolled around like balls and didn't fuck but sprang out of the ground 'like crickets'. Because they were mortal they inevitably began to bother the gods, who retaliated by imposing limits: Zeus took a wire and divided each of them in two, like slicing through a hard-boiled egg. Each human now felt themselves to be partial, half of a whole, a lack that could only be relieved through the intermittent fusion of sex. Fucking – not just heterosexually, and not just for the sake of reproduction, since this was Ancient Greece – temporarily sated the humans' cellular memory of being part of a larger whole.

On seeing the humans delight in their newfound sexuality, Hephaestus observed that each pairing seemed to desire permanent fusion. He laid out his blacksmithing tools and offered his services:

> Do you desire to be joined in the closest possible union, so that you shall not be divided by night or by day? If that is your craving, I am ready to fuse and weld you together in a single piece, that from being two you may be made one; that so long as you live, the pair of you, being as one, may share a single life; and that when you die you may also in Hades yonder be one instead of two, having shared a single death.

Here we are left without an answer, as Aristophanes only fabricated Hephaestus' offer with the intention of proving a point. Aristophanes was convinced that, given the option of being restored to their former wholeness, the humans would accept without a moment's hesitation. In other words, they would become schistosomes. It would be easy to assemble a just-so story in which schistosomes represent the endgame of coupledom, but I think the story would have to be more complicated than that. A pair of schistosomes, it seems to me, is anything but independent from their surroundings, relying on a complex cycle of different species, organisms and environments to maintain their fused position. And any relief arising from their shared life is complicated by the underside of Hephaestus' offer: a shared death.

I learned about the lifecycle of *Schistosoma* worms after spending time in parts of a country where the disease they cause in humans – schistosomiasis or 'snail fever' – is endemic. This was often the way that my fast-paced education in epidemiology went. Symptoms first, causes second. And why wouldn't it? Disease is immediate and connected to an individual, a lived agony that demands alleviation. To study the organisms that cause disease involved temporarily turning my attention away from their role in human suffering. Since I couldn't look with a pragmatic, epidemiological gaze – one which could appreciate the ingenuity of pathogenic organisms while also using my study of them to lessen human suffering – then why was I looking? Was I in some way enjoying sizing up these agents of pain and death?

This is how I knew I could never have been a scientist – it was impossible to get my own anxious ruminating out of the way. Some of the scientists I spent time with seemed to possess what I didn't: something like an internal switch or filter that could separate the task at hand from its wider existential questions and how they themselves might be implicated – at least for the duration of their workday. I knew they had a lot of expertise to share with me, but our conversations often reached an impasse: I had too many unanswerable questions, and the cold hard facts on offer would go over my head or pile up around my ears, unable to sink in: I simply couldn't register. As usual I suspected the problem was me. But then I always found it easy to speak to R—.

R— is an epidemiologist whom I accompanied on several research trips. Unlike the others, I wanted R's approval: I considered them to be serious in a way I could never be, not because they are a scientist (although I was still prone to literary self-abnegation at that point) but because of the way their politics and profession appeared to be seamlessly integrated, each giving purpose to the other. R— was the opposite of a Malthusian. They had gone into epidemiology, specifically diseases of dairy animals, they told me, because they were preoccupied with the question of how humanity would continue to feed itself in the decades to come, soil degradation, epidemic outbreaks and late capitalism considered. Before they were born, their parents lived through a period of political turmoil which resulted in the experience of prolonged food scarcity. Growing up, R— inherited this sense of precarity, an understanding that, in times of crisis, the supply can – and will – without warning, run dry.

R— was interested in the future of food and the genetic modification of crops, in how growing populations would be fed. Their outlook was unfashionable in comparison to the Sunday-supplement food culture I was aware of, which fetishized a regression to the diets of palaeolithic ancestors, or else to the exclusively organic, the all-natural, the soil-grown, fiercely avoiding the nasty artifices of genetic modification. The aesthetic of this culture was enticing – but at what cost? Was there enough grass-fed beef or organic asparagus to feed the world? It's not to say that the two positions are irreconcilable – any good utopian holds space for universal luxury. But still, R—'s opinions on food were refreshing, and realistic. They did not believe that there were too many people at the table of life; they wanted to be able to feed whoever turned up.

We talked about novels and anti-microbial resistance. My notebook was full of vaguely formulated questions like *who is the eater & who gets eaten? who gets to be the food?* I told them what I'd heard from others about disease eradication setting a false precedent, and my concerns about how this position could be easily misconstrued and abused. Saying *adaptation* and *resilience* instead of *eradication* and *control* was one thing, but what, I wanted to know, did this look like in practice? How did it differ meaningfully from the policy of herd immunity that would later be inflicted by the Conservative government in the early stages of the pandemic, which led to thousands of unnecessary deaths among the elderly, disabled and immuno-compromised?

I was grateful to R— for taking my earnestness seriously. They said they couldn't offer an overarching answer, but that in practice the work often involved trying to find loopholes or lay-bys in unexpected places. For example: helminth therapy. Until recently, all intestinal worms were believed to be parasitic on humans, and their near eradication in the Western world was considered a success. But recent studies in autoimmunity seemed to suggest a potentially mutualistic coexistence between worms and humans. R— told me about one of their colleagues who was a lifelong sufferer of asthma, eczema, and other autoimmune problems, all of which had proved unresponsive to treatment. This colleague had heard about the alleged immunomodulating effect of certain worms, and so, while undertaking a field trip to South America, he ingested the larvae of a particular species. This had to be done off-record; getting official clearance from medical channels would have been impossible. Soon the worm grew in his body. The side effects he experienced from being parasitised were the total disappearance of autoimmune symptoms; his life-long asthma and eczema vanished.

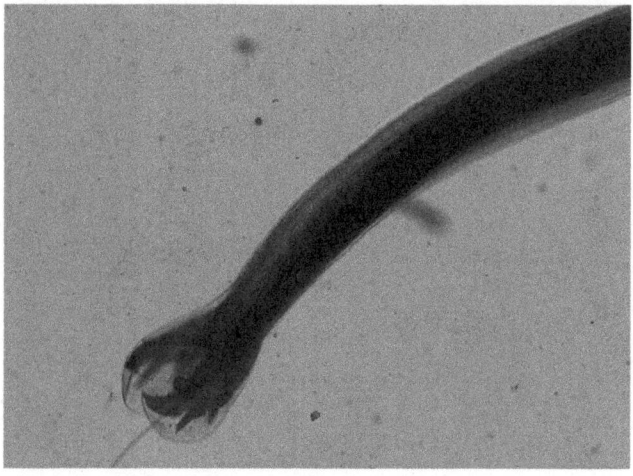

I later read a description of this process in a paper quoting immunologist Mathilde Versini:

> It is worth noting that helminthes [sic] have co-evolved with their host for millennia; their goal is not to kill their host but to survive as long as possible by creating a state of tolerance. To achieve this, helminthes are able, through various mechanisms, to finely modulate the host immune system to prevent an activation that may lead to their elimination, while not causing too deep an immunosuppression which would cause the host to die from infection. This immunomodulation, by avoiding excessive activation of the immune system, contributes to host protection against inflammatory disorders.

Parasitic helminths, then, had located their own loophole of tolerance: it was mutualism nested inside parasitism, an infection that seemed to give both parties what they needed. It sounded a lot, I thought, like love. So I clung to this example for a sense of direction, using it to trot out my issues with apparently clear-cut terms like parasitism, which often turned out to be less straightforward. Much later I realised I was harbouring a fantasy – a hope that all parasitic relationships might turn out, like humans and helminths, to be symbioses that have just been misunderstood, and will one day reveal their mutualistic harmonies. How eschatological of me! It was exactly as my ex said – I was unwilling to accept violence – intentional or otherwise – for what it was. Sometimes an infection might turn out to be a symbiosis in disguise, but sometimes it is just an infection: the multiplication of one organism inside another, resulting in death. It would be hubris to pretend otherwise.

A Species Is An Idea

> *All beings circulate through each other.*
>
> *– Denis Diderot*

On my copy of *Self-Portrait in Green*, Marie NDiaye's author bio begins as follows: 'Marie NDiaye met her father for the first time at age fifteen, two years before publishing her first novel.' When I first read this it wasn't NDiaye's extraordinary precocity that struck me – a novelist at seventeen – so much as the clause that preceded it. I wondered what that admission was doing way out at the front of the paragraph. Typically the author's biography is forced, by the conventions of publishing, into a fiction of its own: a frictionless veneer of professional success, lacquered with as many accolades as possible (there my equanimity slipped, and I typed *personal* instead of *possible*).

So what was the absent father of the Prix Goncourt-winning Marie NDiaye doing in that first sentence?

The child in me thought I understood it, or at least something proximate to it: one thing about first meeting your father at the tail-end of childhood is that it is, among a continent of other feelings, a formative embarrassment that prefaces – and possibly precipitates – any later success. I was ten when I met mine for the first time, and I'm still embarrassed about it. I can only explain this through the trope of the unrequited. It's humiliating to be the seeker of love, rather than the sought. I'd been asking my mum and my grandmother about him for years, trained by the world to conflate origins with truth and meaning. Without access to my origins – him – I'd have neither. Who was he? What was his name? Did he know I existed? And if he did, surely he'd want to meet me, must be as curious about me as I was about him?

The only explanation, I concluded, was that he didn't know I was alive. So the cards were in my hands: I would be the pursuer. I would put on my armour and storm the castle of love. I kept pressing, and details were grudgingly released: a name, the last-known whereabouts. My grandmother's knack for the Yellow Pages. A phone call between him and my grandmother which resulted, at his insistence, in a trip to the GP to test the devotion of my DNA. And then, by the grace of phylogenetic proof, we arranged to meet.

If this was a different kind of book I would describe, in detail, the drive to the town he lived in, the slowness of my grandmother finding a parking place on his street, my hands shaking as we walked along it, counting the house numbers until we reached his door (black, windowless).

I would describe him opening the door, standing next to his wife and son and daughter, and the sequence of events that led up to something he told me later that day, or perhaps it was the following day, or a week later. In any case, it was the first time he and I were in the woods together. We were alone for this part of the conversation, but surely not for long, as it was clear from the outset that the Father was embedded in the Whole, a real nuclear family, which precluded access to him as a lone particle. A crucial detail: my sense of being an interloper in this Whole was historically justified, since my existence resulted from a single night of my father's infidelity. My appearance in this family scene, then, was also the reappearance of his betrayal.

Later that day or week we were walking in the woods and a very large dog ran past and almost knocked me over. I hadn't flinched, but conversation had been awkward prior to the dog and its sudden appearance gave my dad a prompt. Did I know what to do, he asked, in the event of being attacked by a dog? I didn't.

He told me that a dog would typically attack by leaping through the air, front legs first, hoping to knock and pin me to the ground, where it could access my neck with its teeth. And by that point, if the dog was stronger than me, it would already be too late. The first few seconds are crucial, he said, so I'm going to tell you what to do. Okay, I said, ready for my first dose of fatherly advice. When the dog is in the air coming towards you, he said, you need to grab onto its front legs like this – he raised two hands in the air, closing them into fists around the invisible legs – and PULL. He yanked his fists away from each other like he was snapping a giant wishbone. A dog's heart, he told me, is located directly in the middle of its chest, and so pulling the front legs apart will cause an immediate rupture: the dog will be dead before it hits the ground.

I remember being impressed by this – or perhaps I remember feeling like I *should* be impressed, or grateful, but there was a hot shameful absence where those feelings should be. It's embarrassing to have to feign enthrallment when you are actually disappointed, when the object of your seeking – now found – clearly didn't want to be sought. You can't cover up what you've forced into light. And what will you do, now there's nothing left to seek?

○

When I learned about the origins of viruses, my initial response was absurd recognition. *I know how you feel!* I wanted to say back to the textbooks I was reading. According to the conventional definition of what constitutes life, viruses are outsiders, marginal entities in the drift-space between life and lifelessness. Every other organism on earth can be defined by having a cell (or cells) of one's own, but viruses lack this fundamental architecture that would permit entry into verified life. Since they are indeterminate bundles of genetic material, viruses must seek out and override other cells in order to function. If a cell is a city, as the so-called father of pathology Rudolf Virchow conceived of it, then viruses are the shadows that move beyond the city walls, peripheral and aberrant.

Theories of how viruses came to exist are similarly indeterminate. They could be 'mobile genetic elements' that somehow developed the ability to jump from cell to cell; they could be the descendants of organisms that adapted a parasitic replication strategy. It could be that viruses are what came before cells, giving rise to cellular life as we know it. Lynn Margulis describes the 'escape hypothesis', in which viruses started out as nucleic acid inside a cell, somehow leaked out, then could not get back in. This exiled genetic material then tried to re-enter every cell it encountered. The escape hypothesis figures viruses as prodigal matter, stranded genetic material just trying to get home.

I'm afraid to say that this is the description that pulled my overidentification strings. I knew what it was like to be an unintentional parasite with troubled origins, to rupture another's perceived harmony for want of belonging. It was compelling to recast the nasty virus tearing through cell after cell as some Homeric protagonist, passing through many worlds in the attempt to reach familiar shores.

Or maybe, I backtracked, viruses don't want to get home – maybe they rejected the universal architecture of cells and thought they could do better without. What's so great about a cell, anyway? It has an inside and an outside, and a cell wall between. This border is a necessary limit, but it's also a membrane that enables contact with other cells. It reminds me of Simone Weil's observations on the 'metaxu', a metaphysical state of contradiction and mediation, which she likens to a wall between two prisoners:

> Two prisoners whose cells adjoin communicate with each other by knocking on the wall. The wall is the thing which separates them but it is also their means

of communication. It is the same with us and God. Every separation is a link.

That final clause might be the catchphrase of cellular existence. Like the metaxu, a cell is an elegant paradox, since it is both what separates and what keeps us together. But being in touch, like the prisoners Weil describes, doesn't mean that we will like everything we hear or receive from the other side.

I might be *like* a virus, but at least I could console myself that I wasn't actually viral matter, since unlike a virus I had cells of my own. This was complicated by reading that, genetically speaking, human viruses are probably more closely related to humans than they are to viruses that don't target human cells. I may think that viruses are in one category and humans are in another, but my DNA would suggest otherwise: genetically, the polio virus may have more in common with me than it does with the mosaic virus of tobacco plants.

For medical anthropologist David Napier, this blurred boundary between human and virus troubles the Enlightenment model of 'selfhood' that figures the self as discrete and unchanging, clearly distinguishable from other beings and from its environment. Viruses upset this by being both self and non-self at the same time. They are considered non-living, but at the same time, the way they are described in scientific literature poses a contradiction. If viruses are not 'alive', Napier asks,

> How can viruses *recognize, scout, trick, discover, alert, evade, sense, recruit, mobilize, prod, mask, defend, scavenge,*

attack, invade, adapt, appropriate, sacrifice and *kill* if they lack mobility and *do not respond to environmental stimuli*?

I like how this becomes a question of grammar, if not poetics: verbs are a sign of life, not lifelessness. If a virus isn't alive, whose liveliness animates it, other than my own? And if I am the one animating it, does that mean that the virus 'attacking' me is a part of me?

○

Autoimmunity scholars have long grappled with the existential slipperiness of bodies that are perceived to work against themselves. Ed Cohen – who was diagnosed with Crohn's disease at the age of thirteen – recounts how the doctors who tried to explain his condition to him likened autoimmunity to 'being allergic to yourself', or 'eating yourself alive'. Donna Haraway also locates a shared metaphorical paradigm between the pathogens that might invade us, and the stigma of autoimmunity:

> We seem invaded not just by the threatening 'non-selves' that the immune system guards against, but more fundamentally by our own strange parts. No wonder auto-immune disease carries such awful significance, marked from the first suspicion of its existence in 1901 by Morgenroth and Erhlich's term, *horror autotoxicus*.

Growing up under the sign of such precarious metaphors led Cohen to wonder about the definition of 'self', and the extent to which his own was considered a deviation. Cohen charts how, since its inception at the end of the nineteenth century, immunology has absorbed ideas of selfhood from Enlightenment thinkers like John Locke, who located man's identity 'in one fitly organised body'. This is a model in which the self is to the body what a property owner is to his estate. But autoimmunity – and, following Napier, viruses – complicate this assumption. If the body cannot consistently distinguish between self and non-self, Cohen asks, 'how stable can the received notions of "self" be?'.

I would like to feel coherent, like I am on my own team, but physically and psychologically this is rarely the case. Something in the self is always tripping itself up, fucking around and finding out, an unending lovers' quarrel between sabotage and survival.

Some days – like the halved-humans in Aristophanes' speech – I fantasise that I might be fixed by the reinstatement of some mysterious Thing I am lacking. This Thing would make me whole and coherent, a full self. When I was younger and less well-versed in the ways of the Thing it was easy to conflate it with the individual people it sometimes appeared to settle on or around, like a radiant cloud. Not understanding that the Thing was amorphous and errant by nature, I would become absorbed in the person in the hopes of finally communing with the Thing and attaining a full, real self. These attempts were often fleeting – sickly-sweet or swiftly bitter. The Thing's powers of illusion can be so convincing that I might never get it into my skull that no one person can be contiguous with it. Its most graspable form is comparable to sillage: an imprint of perfume that lingers in the air long after the wearer has left the room. Necessarily absent, the Thing is a tantalising quotation of presence.

I encountered something proximate to this on first reading *Powers of Horror*. Kristeva defines abjection as the violence of mourning for something that has always already been lost. This something, the object, is only discernible by its felt lack, the sense that something vital is missing from the self. The lack might manifest as phobia, or a violent, fearful conviction that the self is inherently wrong or contaminated, suffused with something *bad* that must, somehow, be cast out. Kristeva locates this lost object very early on in life, describing a child who, before the acquisition of language, 'has swallowed up his parents too soon'. As a result, the parents can no longer safely shape the world for the child, they can no longer form the sky and firm ground between which he lives. The child has internalised a terrifying, edgeless universe, and so in order to save himself from it he rejects everything; 'Fear cements his compound.'

In *Powers of Horror*, abjection is based on a principle of exclusion: what must be rejected and cast out for the self to feel safe. The origins of my own childhood phobia were typically banal: a school nature study aged six, a slice of cheap white bread. We were taken to the edge of the playground, which bled into scuzzy urban woodland, and instructed to bury our slices a few centimetres below the surface of the soil. A week later we exhumed the bread and drew pictures of all the bugs and worms that had made their 'home' in the slices. For the next fifteen years I could not look at a slice of white bread without gagging – friends' parents were impressed if unconvinced by my preference for brown bread; it seemed pretentious, an affectation.

At the same time, I became obsessed by what had infested the bread. At school I collected ladybirds, worms and – my favourite – woodlice, and spent hours creating what I imagined to be ideal environments for them inside plastic boxes: a jungle of twigs and leaves, a frosted cornflake for food and a pink barbie stiletto filled with water. I was moving schools and homes every year or few years, but I have a bug-related memory attached to each of them: the ecstasy of seeing a tiny blizzard of louse eggs hatch in my cupped palm; being hypnotised by the brightly-coloured fishing maggots 'racing' at a school fair; the white cloud of mucus forming between two black slugs, as if one was sicking up the other (I've since found out they were mating). The bugs I could stomach, but if you had offered me, on any one of these occasions, a white bread sandwich – I might have thrown up.

An object or substance does not have to be manifestly defiling to invoke horror in the beholder. As poet and psychoanalyst Nuar Alsadir puts it, a phobia functions like a joke or a dream, 'which is not the source of unsettling emotion but its container, the vehicle through which emotions like aggression, sexual desire, and other id impulses flagged by the superego can be avoided.' Sometimes the substance that would appear to lend itself to phobic containment becomes a source of attraction, like my displacement of bugs for bread, a tainted Eucharist.

Georges Bataille's novella *Blue of Noon* can be read, I think, as an ode to this kind of covert abjection. Its surface preoccupation is evident: soaked on every other page with piss or vomit, and a Bataille-like narrator who descends into illness while oscillating between two young women, Dirty and Lazare. Dirty is an animal-eyed self-debaser who can frequently be found in pools of her own urine or hysterically retelling the death of her mother. Lazare, by contrast, is a self-serious political revolutionary, acknowledged to be a fictional rendering of Simone Weil. The overt abjection of Dirty, I think, is meant to act as a foil for the covert abjection of Lazare, whose name echoes the disturbed grave of Lazarus, or the lazaretto plague islands where ships carrying sickness were quarantined.

This checks out with the narrator's growing obsession with Lazare. He is disgusted by her moral cleanness and admits that 'Lazare revolts me so much that it scares me'. He is revolted, not by actual filth or his abject lover Dirty, but by the obscenity he perceives at the heart of Lazare's purity. Phobia is in the eye of the beholder after all, or, as anthropologist Mary Douglas famously put it, 'dirt is matter out of place'. While Lazare might not seem immediately abject, *Powers of Horror* reminds me that it is often 'not lack of cleanliness or health that causes abjection but what disturbs identity, system, order.' Lazare disturbs the narrator through his conviction that she harbours, as her name suggests, a secret love of death. The scholar Alexander Irwin has commented:

> With his depiction of Weil as Lazare, Bataille wants to proclaim that the image of the life-loving revolutionary

is based on self-deception. Those who proclaim the ideology of life nourish a secret necrophilia.

Bataille's treatment of Weil's politics through the character of Lazare reaches for abjection's ungraspable essence, back through life, right into the grave itself. Reading *Blue of Noon* I am often unable, bad reader that I am, to isolate the narrator and Lazare as characters within the text. They are also Bataille and Weil, locked in a clash of symbols: the undying Acéphale whose head may have been devoured by Weil the cannibal, who is herself being parasited by God and bacteria, which may, in the illogic of digestive mysticism, be one and the same thing.

○

Only in the final stages of re-writing this book did I make the link between its subject matter and my childhood love of bugs. As a teenager I mostly forgot about them, swatting them away while I revised for exams, and ditching science as soon as I was able to. I was eating white bread by the time I got to art school, although only in small doses. I was busy trying to grow up, to forget an ugly childhood body that remembered worms, warts, poxes, rashes. Childhood was the night-knowledge of being indeterminate matter, the body as an ellipsis, a page on which anything might be written or deposited. So I didn't think about bugs, wasn't especially hygienic – minor infections came and went. They didn't return to the foreground of my mind until I started my research project. And then, along with the delight came the horror – no longer for bread, but for my capacity to be invisibly infected.

First there were months of being around actual pathogenic samples in labs, handling petri dishes of *Salmonella* and *E. coli*, knowing anthrax was kept in the next room. But I never connected these things to my body, to what might erupt from their encounter. Later, unable to do field research during the pandemic, I bought a microscope and started to collect water samples from ponds and puddles and drainpipes. I'd never done this kind of thing as a child, but once I started, I began to understand the seventeenth-century mania for 'animalcules': I spent hours on water microscopy message boards, trying to determine my *Paramecia* from my *Vorticella*, my rotifers from my tardigrades. It was like how a friend had described the sudden trend for mushroom identification – Pokémon for grownups. I sometimes had a passing thought of how many of these things I had swallowed in my lifetime – along with their dense habitat of algae and diatoms – but the Ick didn't arrive until I accidentally grew a colony of *Colpidium colpoda* in a bucket of rotting flowers.

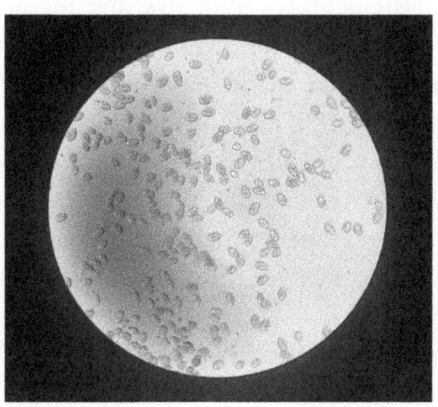

One day I checked a drop of this bucket water under the microscope and saw a few kidney-bean-shaped things zooming around, varying in colour from translucent to pale yellow. On closer magnification I could see the stained-glass windows of their insides, the intricate organelles. When I checked again in the evening there were swarms of them. By morning the drop of water appeared to be gone and in its place was one pulsing organ of matter; the organisms had bloomed into every available space until their sum exceeded their parts. That's when my stomach began to churn. *Colpidium colpoda*, to be clear, are not pathogenic. But they do divide quickly, every four to six hours, especially when feeding on a plentiful bacterial soup, which the decomposing flowers had unintentionally provided. My message board scrolling, once pleasurable, was now panicked – my guts and I were convinced that I had ingested some of these organisms, that some nasty infection, some water-borne disease, was now brewing inside us.

Once the grip of this panic released me, I looked differently at Pasteur's table manners, and the germophobia that also consumed Simone Weil as a child. It's terrifying to feel so open to the world – as if looking – even thinking – might really be a kind of ingestion. No wonder the discourse of immunity, built on Pasteur's legacy, became all about the fortress self, a buttressing of borders and strongholds. Fear cements the compound, as Kristeva put it. But if abjection is a dance with death, it also generates life and creativity.

In order to be lived with, the abject must be purified through catharsis. Kristeva describes how this used to take place through religion and ritual but is now primarily carried out through the act of creativity, which is 'rooted in the abject it utters and by the same token purifies.'

According to this theory, abjection is a state 'eminently productive of culture' which seeks to bind an unbound world in signs and language. I have my childhood phobia to thank, maybe, for the existence of this book. But if its writing became a kind of catharsis, it's certainly not one that leaves abjection behind.

I find this push-pull between defence and creativity – the making and unmaking of selfhood – useful for thinking about immunity. If abjection is the mourning for an object always already lost, the immunological correlate to this lost object might be the myth of a fortress self, invulnerable to infectious others and its own mutinous parts. It's always easier, after all, to split off your own aggression into an external agent. When the dog nearly knocked me over and my dad told me how to kill it, this helped to consolidate our familial bond, united against external threat. But I've since wondered if it was something other than the dog that prompted his talk of killing. I, after all, was the beast that invaded his family from the outside, laying siege to its fortress, a stray animal threatening the Oedipal order. But he couldn't kill me, any more than he could kill the mutinous parts of himself that had produced me.

It's always a relief, then, when a real dog comes along. Or a real bug – those perfect containers for murderous wishes. My fear of death from a bucket of water was in some ways a welcome escape from my other preoccupation at the time – the legal and financial hell I was over a year into and could not see a way of surviving. Someone wanted me dead, it felt like, although it probably wasn't the *Colpidium colpoda*. And probably I wanted someone dead too: my dreams, for the first time, were of myself as a killer rather than the one being killed. This reversal dispatched with my short-lived phobia, as well as the *Colpidium* colony itself, when I viciously emptied the bucket down the drain.

◌

But if immunity has something to do with abjection, where does this leave the resulting creativity that Kristeva describes? The functioning of the immune system doesn't immediately suggest a 'creative' process, other than the antibody it creates when it senses a pathogenic intruder. Later, I read something that broke my received understanding of immunology apart. Rather than just making antibodies that match intruders, the human body creates millions of variations of antibodies, prior to any detected threat, the majority of which are never used. This is known as an evolutionary paradox because according to the laws of natural selection it's a colossal waste of energy; numbers of variations should be limited to matching actual threats, rather than speculative false alarms. It's wasteful and pointless, according to Darwinian logic, to create millions of antibodies that never get used. It's inefficient by the standards of capital too, since it denies the scarcity that forces subjects into endless competition and productivity.

But, as Napier points out, this is only a paradox if we rely on the assumption that wild variability within the body is an aberration, rather than the norm, and that self-defence is the only reason for mutation and variation. He suggests a different logic for this abundance of antibodies: a model of selfhood that is in constant, curious engagement with its environment. One that is less interested in shoring up what it already knows about itself, and more interested in engaging with the difference it finds in its encounters. Some of these encounters might prove to be harmful, but others will change the course of life as we know it – like the mammalian virus that enabled placenta to become symbiotic, rather than parasitic – becoming part of what we now think of as ourselves. It's immunity as radical vulnerability, rather than pure defensiveness, a push-pull – like Weil's metaxu – between what we can learn to accommodate, and what we cannot.

This logic of flourishing and abundance is not without risks. It reminds me of what Bataille would call a limit-experience, in which the self encounters and is irrevocably changed by a seemingly impossible, unliveable intensity or otherness. The limit-experience is ecstasy mixed in with horror, life with death. It might manifest through religious or mystical encounter, an illness or near-death experience, or the quotidian insanity of falling in love.

Italo Calvino's story 'Mitosis' traces the limit-experience of bacterial life. Like many stories in Calvino's *Cosmicomics* series it's narrated by the recurring, amorphous character of Qfwfq. Here Qfwfq is a cell, who begins by telling us he is 'dying of love', then immediately negates the idea that love implies the existence of another.

Rather, being himself is all he knows, with no concept of an outside or another being separate to himself. Qfwfq is saturated in the umwelt of himself to the extent that he brushes up against and discovers his own boundaries, awakening the possibility of an outside: 'I had this contentment because outside of me there was this void that wasn't me, which perhaps could become me because "me" was the only word I knew'.

Qfwfq experiences a burgeoning awareness of the space outside him, and, like a mystic seeking to escape themselves in ecstasy, this space quickly becomes the locus of desire. Desire creates a budding dialectic between outside and inside, which can only be traversed through the expression of that desire. Qfwfq tells us he is moved to express himself, but being a single-celled organism all he has to hand is the language of his own being: the genetic material of the nucleus. So he expresses this genetic material, and then he expresses it again and again. This refrain seems to presage Gertrude Stein's understanding that repetition never creates sameness but gradations of difference, since by repeating himself Qfwfq replicates his chromosomes ready for mitotic division. Through repetition he produces a state different to the one in which he began.

When Qfwfq expresses the desire to get outside of himself – ecstasy – he creates the possibility of another within himself, which through mitosis manifests as another outside of himself. Or rather, he is dispersed over the two. In Calvino's narrative, mitosis is depicted as more than the mindless cloning of minimally sentient life; it is the beginning of a cosmic love story whose protagonists are propelled by some vital, intangible force.

But what if Qfwfq is not an anonymous cell, or Freud's narcissistic amoeba, or a harmless *Colpidium colpoda* feeding on rotten flowers – but a pathogenic bacterium inside the body of a stranger, or in someone I love? What if the love story of Qfwfq's limit-experience is simultaneously the story of my chronic infection, or the tubercular illness that proved fatal to Simone Weil? *Mycobacterium tuberculosis* is an appropriate candidate for Qfwfq's slow and sensual cell division, since compared to other bacteria it practices an almost lethargic rate of division, undergoing mitosis just once a day. When an immune system recognises mycobacterial cells it enlists white blood cells which attempt to engulf them. But the composition of the mycobacterial cells enables them to resist this engulfing, and they continue to multiply in the body. The mitosis of mycobacteria gives rise to the potentially fatal human disease, tuberculosis; Qfwfq's ecstasy multiplies unchecked.

Here is the cellular version of digestive mysticism – there's no living or dying without these conflicting points of engulfing, digesting, merging and coming apart. It's like encountering a multispecies depressive position, where life and death are streaming from the same source. Infections, whether or not I want them to, pluck at my symbiotic heartstrings, reminding me that we are both eaters and eaten.

My own is in a period of dormancy: no longer severe enough to require intravenous antibiotics, but never absent enough for long enough to let me forget it is there. Sometimes I feel calmly disposed to its presence, and at other times panicked – as if its apparent benignity is a deliberate ruse, there to distract me while it seeds into malignance. It reminds me of the remark made by Gillian

Rose's doctor towards the end of *Love's Work*. 'You are living in symbiosis with the disease,' he tells her; 'go away and continue to do so.' Rose, it transpired, did not have long to endure this symbiosis. The difficulty of limbo is underrated, I think, and perhaps this is why *Love's Work* unravels from the counsel of its epigraph – 'Keep your mind in hell, and despair not.' Like Rose, I also want to keep my mind in hell; since hell, as I learned as a child from the prophet Isaiah, is a place where worms do not die.

I tried to make it okay but I couldn't. The parasitism never became a symbiosis, there was damage instead of reconciliation. My inbox sang of legal threats, and backlogs of rent and bills. I walked into the woods where the birds were ringing the rain into stillness. There were drops on all the leaves as if poised ammunition. I was *looking for something* but what. It was almost like I am *looking for something to hurt* but not quite that either, since I knew the urge to hurt and kill was a new way of grieving the older *looking*, which was always a quest for the Thing, the renewable hope of a virus or of a child.

I looked at a tree, rotted and felled, choked all around with ivy. I thought: *all attachment is optimistic*. This was not the Thing but my body seemed to think it would do for our purposes, and soon I was the killer and the ivy was the damage and the tree was already dead but I was kicking it free. There was a hotness going through me and if this hotness had a noun it would be *anger* but if it had a verb it would be *exorcising*. The ivy was binding the tree like a demon around a spine and I was kicking the damage from its outline, the future from its origins, loosing matter back into the world. It went into the ground and the air and the places I couldn't see, none of which were immune to its pursuit of wild contradiction. To be on the side of life you must sometimes be antibiotic.

Acknowledgements

At the University of Glasgow: Professors Zoë Strachan, Jo Sharp and Ruth Zadoks for helping me secure an interdisciplinary scholarship, and for guidance and important conversations throughout the PhD. Jennika Virhia, Alicia Davis, Professor Sarah Cleaveland, and everyone involved in the zoonoses projects who gave me time and patience to engage with their research. Professors Lavinia Greenlaw and Felicity Callard for examining an earlier version of this text for my viva, and for the immense generosity of attention and feedback that encouraged me to pursue it further.

Patrick Farmer for believing I could write this book, and for love, inspiration and company while I completed a first draft.

Gwen Dupré for digestive mystique and Simone Weil insights. Kirsty Hendry for telling me about *bugge*. Sarah Shin for sharing articles about amorous microbes. Mike Saunders for telling me to watch *Upstream Colour*. Maria Sledmere for reminding me about the antibodies in *Malina*. Will Rees for directing me back to Freud's amoeba.

Friends whose insights in the final editing stages made this book better, and whose encouragement saved me from total dejection: Sarah Bernstein, Helen Charman,

Oli Hazzard, Sam Keogh, Christopher Law, Lucy Mercer, Will Rees and Niall Tessier-Lavigne.

Editors and programmers at The White Review, Wellcome Collection, Serpentine Galleries, The University of Edinburgh Art Collection, Ma Bibliothèque, The Drouth, Granta, Daunt Books Publishing and Bricks From The Kiln, where some earlier iterations of this research first appeared.

Jake Franklin, Sam Fisher and Will Rees at Peninsula Press for welcoming this strange book. Karolina Sutton for her ongoing support and patience.

Sam and Uisce for love with unnerving constancy.

Images

p5: Detail from Ms. Rh. 172 – Aurora consurgens / f. 19v-40. Reproduced courtesy Zentralbibliothek, Zürich.

p31: Detail from coffret lid 'Minnekästchen' c. 1325-50, Upper Rhineland, Germany. Rogers Fund and The Cloisters Collection, by exchange, 1950. In public domain.

p49: Drawing by 10-year-old Richard, one of Klein's patients, 1941. Reproduced courtesy The Melanie Klein Trust.

p61: *Metamorphoses* Chapter p69. Sarah Craske, 2016-2017. Reproduced courtesy the artist.

p75: André Masson's cover for the first issue of Acéphale, 1936. Reproduced courtesy University of Missouri.

p81: Origen emasculating himself. Detail from Roman de la Rose, c. 15th century, France. Bodleian Library, MS. Douce 195, fol. 122v. In public domain.

p87: 'What are you?' 'A worker. And we outnumber you.' Meme, anonymous, undated.

p94: Leucochloridium paradoxum; parasite in Succinea putris. Monographie des succinées francaises, Auguste Adolphe Baudon, 1879. In public domain.

p105: The Eucharistic Man of Sorrows, Friedrich Herlin. 1469, Germany. Reproduced courtesy Stadtmuseum Nördlingen.

p113: Bdelloid rotifer. 11 February 1977. Ian Sutton, Oberon, Australia. Reproduced courtesy Creative Commons.

p115: Couple of Schistosoma mansoni. 27 December 2016. Alaa. Reproduced courtesy Creative Commons.

p121: Necator americanus. 25 April 2018. Daniel J Drew, Yale Peabody Museum. Reproduced courtesy Creative Commons.

p137: Colpidium colpoda. 4 June 2020. Daisy Lafarge, Sheffield, UK. Reproduced courtesy the author.

References

Page nine quotes from 'Love (III)' by George Herbert, and page one hundred forty-seven from Lauren Berlant's *Cruel Optimism*. All other references can be found in the following, directly or indirectly:

Jill Haak Adels ed., *The Wisdom of The Saints: An Anthology*
Hajar Ahmed Hajar Albinali, "Majnoon Lila"
Nuar Alsadir, *Animal Joy*
G. L. Ataev et al., "Multiple infection of amber Succinea putris snails with sporocysts of Leucochloridium spp. (Trematoda)"
Teresa of Ávila, *The Interior Castle*
Ingeborg Bachmann, *Malina*
Georges Bataille, *Blue of Noon*
Georges Bataille, *Literature and Evil*
Georges Bataille, *The Sacred Conspiracy*
Georges Bataille, *Visions of Excess: Selected Writings, 1927-1929*
Gregory Bateson, *Steps to an Ecology of Mind*
Michael Begon et al., *Ecology: Individuals, Populations and Communities*
Walter Benjamin, *Reflections: Essays, Aphorisms, Autobiographical Writings*
Rae Ellen Bichell and Melody Schreiber "Bubonic Plague Strikes In Mongolia: Why Is It Still A Threat?"
Eula Biss, *On Immunity: An Inoculation*
William Blake, *Songs of Innocence and of Experience*
Giovanni Boccaccio, *The Decameron*
Julia Borossa, ed. *Sándor Ferenczi: Selected Writings*
Bruce Braun, "Biopolitics and the molecularization of life"
Norman O. Brown, *Love's Body*

Sir Thomas Browne, *Selected Writings*
Albert O. Bush et al., *Parasitism: The Diversity and Ecology of Animal Parasites*
Caroline Walker Bynum, *Holy Feast and Holy Fast: The Religious Significance of Food to Medieval Women*
Italo Calvino, *Cosmicomics*
Italo Calvino, *Six Memos for the Next Millennium*
Italo Calvino, *t zero*
Michael Camille, *The Medieval Art of Love: Objects and Subjects of Desire*
Daniel F. Caner, "The Practice and Prohibition of Self-Castration in Early Christianity"
John Carey ed., *The Faber Book of Science*
Shane Carruth, *Upstream Colour*
Anne Carson, *Eros the Bittersweet*
Hélène Cixous, *Stigmata: Escaping Texts*
Emanuele Coccia "All Species Have the Same Life"
Ed Cohen, *A Body Worth Defending: Immunity, Biopolitics and the Apotheosis of the Modern Body*
Ed Cohen, "My self as an other: on autoimmunity and "other" paradoxes"
Alicia Davis and Jo Sharp, "Rethinking One Health: Emergent human, animal and environmental assemblages"
Vivian Delchamps, ""The Names of Sickness": Emily Dickinson, Diagnostic Reading, and Articulating Disability"
Gilles Deleuze and Felix Guattari, *A Thousand Plateaus: Capitalism and Schizophrenia*
John J. Dennehy, "Evolutionary ecology of virus emergence"
Robert Desowitz, *Tropical Diseases: From 50,000 BC to 2500 AD*
Les Dethlefsen et al., "An ecological and evolutionary perspective on human-microbe mutualism and disease"
Emily Dickinson, *The Complete Poems*
Denis Diderot, *Rameau's Nephew/ D'Alembert's Dream*

Kimberley A. Dill-McFarland, et al., "Close social relationships correlate with human gut microbiota composition"

Annie Dillard, *Holy The Firm*

Clifford Dobell ed. *Antony van Leewenheok and his "Little Animals"*

John Donne, *A Selection of His Poetry by John Hayward*

Mary Douglas, *Purity and Danger: An Analysis of Concepts of Pollution and Taboo*

Rene J. Dubos, *Louis Pasteur: Free Lance of Science*

John David Ebert ed.. *Twilight of the Clockwork God: Conversations on Science and Spirituality at the End of an Age*

Veronica Forrest-Thomson, *Collected Poems*

Gavin Francis, "The Untreatable"

Paulo Freire, *Pedagogy of the Oppressed*

Sigmund Freud, *An Outline of Psychoanalysis*

Sigmund Freud, *A Phylogenetic Fantasy: Overview of the Transference Neuroses*

Martin Gardner ed. *The Sacred Beetle and other great essays in science*

Christina Gerhardt, "The Ethics of Animals in Adorno and Kafka"

Scott F. Gilbert et al., "A Symbiotic View of Life: We Have Never Been Individuals"

Franklin Ginn et al., "Flourishing with Awkward Creatures: Togetherness, Vulnerability, Killing"

Nahum N. Glatzer ed., *The Complete Short Stories of Franz Kafka*

Édouard Glissant, *Poetics of Relation*

Jared Haft Goldstein "Darwin, Chagas', Mind and Body"

J. B. S. Haldane, "Disease and evolution"

Emilie Griffin ed. *Hildegard of Bingen: Selections from Her Writings*

Hervé Guibert, *Paradise*

Donna J. Haraway, *Simians, Cyborgs, and Women: The Reinvention of Nature*

Donna J. Haraway, *Staying with the Trouble: Making Kin in the Chthulucene*

Donna J. Haraway, *The Companion Species Manifesto: Dogs, People, and Significant Otherness*
Donna J. Haraway, *When Species Meet*
Sandra Harding, *Sciences from Below: Feminisms, Postcolonialities, and Modernisms*
Saidiya Hartman, *Scenes of Subjection: Terror, Slavery and Self-Making In Nineteenth Century America*
Lyn Hejinian, *The Language of Inquiry*
Stefan Helmreich, "Homo Microbis: The Human Microbiome, Figural, Literal, Political"
Steve Hinchcliffe et al., "Biosecurity and the topologies of infected life: From borderlines to borderlands"
Myra J. Hird, *The Origins of Sociable Life: Evolution After Science Studies*
Susan Howe, *My Emily Dickinson*
Alexander Irwin, "Devoured by God: Cannibalism, Mysticism and Ethics in Simone Weil"
Alexander Irwin, *Saints of the Impossible: Bataille, Weil, and the Politics of the Sacred*
Walter Kaufman ed., *The Portable Nietzsche*
Maeve Kennedy, "Bacteria from 300-year-old Ovid poetry volume inspires 'bio-artist'"
Melanie Klein, *Narrative of a Child Analysis*
Julia Kristeva, *Melanie Klein*
Julia Kristeva, *Powers of Horror: An Essay on Abjection*
George Lakoff and Mark Johnson, *Metaphors We Live By*
Darian Leader and David Corfield, *Why Do People Get Ill?*
Thomas Lekan, "A Natural History of Modernity: Bernhard Grzimek and the Globalization of Environmental *Kulturkritik*"
Geoffrey Lapage, *Animals Parasitic in Man*
Bruno Latour, *Pandora's Hope: Essays on the Reality of Science Studies*
Bruno Latour, *The Pasteurization of France*
Bruno Latour, *We Have Never Been Modern*

John Locke, *Essay Concerning Human Understanding*

Jamie Lorimer, "Gut Buddies: Multispecies Studies and the Microbiome"

Sofia Lourenço and Jorge Mestre Palmeirim, "Which factors regulate the reproduction of ectoparasites of temperate-zone cave-dwelling bats?"

Sarah Maitland and Wendy Mulford, *Virtuous Magic: Women Saints and Their Meanings*

Michael Marder, "The Life of Plants and the Limits of Empathy"

Lynn Margulis, *Symbiotic Planet: A New Look at Evolution*

Lynn Margulis and René Fester ed. *Symbiosis as a Source of Evolutionary Innovation: Speciation and Morphogenesis*

Lynn Margulis and Dorion Sagan, *Mystery Dance: On the Evolution of Human Sexuality*

Lynn Margulis and Dorion Sagan, *What is Life?*

Lynn Margulis and Dorion Sagan, *What is Sex?*

Lynn Margulis and Karlene V. Schwartz, *Five Kingdoms: An Illustrated Guide to the Phyla of Life on Earth*

Emily Martin, "The Egg and The Sperm: How Science Has Constructed a Romance Based on Stereotypical Male-Female Roles"

Cristina Mazzoni, *The Women in God's Kitchen: Cooking, Eating and Spiritual Writing*

Margaret McFall-Ngai, "Noticing Microbial Worlds: The Postmodern Synthesis in Biology"

Margaret McFall-Ngai et al., "Animals in a bacterial world, a new imperative for the life sciences"

David McLellan, *Simone Weil: Utopian Pessimist*

Jacques-Alain Miller ed., *The Seminar of Jacques Lacan, Book I*

Trinh T. Minh-ha, *Woman, Native, Other: Writing Postcoloniality and Feminism*

Juliet Mitchell ed., *The Selected Melanie Klein*

Piers D. Mitchell et al. "The Intestinal Parasites of King Richard III"

Timothy Morton, *Dark Ecology: For a Logic of Future Coexistence*
Penelope Murray ed. *Classical Literary Criticism*
Alex M. Nading, "Humans, Animals and Health: From Ecology to Entanglement"
David A. Napier "Nonself Help: How Immunology Might Reframe the Enlightenment"
Marie NDiaye, *Self Portrait In Green*
Maggie Nelson, *Bluets*
Barbara Newman, *God and the Goddesses: Vision, Poetry and Belief in the Middle Ages*
Jacques Nicolle, *Louis Pasteur: A Master of Scientific Enquiry*
Lorine Niedecker, *Collected Works*
Ovid, *Metamorphoses*
Cynthia Ozick, *Metaphor & Memory*
Heather Paxson, "Post-Pasteurian Cultures: The Microbiopolitics of raw milk cheese in the United States"
Hugh Pennington, *When Food Kills: BSE, E.coli and Disaster Science*
Plato, *The Symposium*
Richard Preston, "Crisis in the Hot Zone"
David Quammen, *Spillover: Animal Infections and the Next Human Pandemic*
Gerald Raunig, *Dividuum: Machinic Capitalism and Molecular Revolution Vol. 1.*
Denise Riley, *Impersonal Passion: Language as Affect*
Denise Riley, *The Words of Selves: Identification, Solidarity, Irony*
Lisa Robertson, *Anemones: A Simone Weil Project*
Lisa Robertson, *Nilling: Prose Essays on Noise, Pornography, The Codex, Melancholy, Lucretius, Folds, Cities and Related Aporias*
Gillian Rose, *Love's Work*
Jacqueline Rose, "To die one's own death – thinking with Sigmund Freud in a time of pandemic"
Jean-Jacques Rousseau, *Confessions*
Dorion Sagan ed. *Lynn Margulis: The Life and Legacy of a Scientific Rebel*

James C. Scott, *Against the Grain: A Deep History of the Earliest States*
Eve Kosofsky Sedgwick, *Touching Feeling: Affect, Pedagogy, Performativity*
Michael A. Sells, *Mystical Languages of Unsaying*
Michel Serres, *The Birth of Physics*
Michel Serres, *The Parasite*
Michel Serres and Bruno Latour, *Conversations on Science, Culture, and Time*
Idries Shah, *The Sufis*
William Shakespeare, *The Tempest*
William Shakespeare, *Twelfth Night*
William Shakespeare, *Hamlet*
Steven Shapin, "A Pox on the Poor"
Eleni Sikelianos, "Mitosis, Meiosis, Poeisis"
Eleni Sikelianos, "Refuse/Refuge: Be Longing"
Himali Singh Soin and Tyler Rai, "Safe Travels"
Susan Sontag, *Illness as Metaphor*
Susan Sontag, *AIDS and its Metaphors*
Susan Sontag, *Reborn: Journals & Notebooks 1947-1963*
Jean Starobinski, *A History of Medicine*
Michael Taussig, *Mimesis and Alterity: A Particular History of the Senses*
Jakob von Uexküll, *A Foray into the Worlds of Animals and Humans*
Christian Urs, "Molecular Mimicry"
Joanna Verran and Xavier Aldana Reyes, "Emerging Infectious Literatures and the Zombie Condition"
Luis P. Villarreal, "Are Viruses Alive?"
Tyler Volk, "Seeing Deeper into Gaia Theory: A Reply to Lovelock's Response"
Priscilla Wald, *Contagious: Cultures, Carriers, and the Outbreak Narrative*
Simone Weil, *First and Last Notebooks*
Simone Weil *Gravity and Grace*

D. R. Wessner, "The Origins of Viruses"
Wendy Wheeler, *The Whole Creature: Complexity, Biosemiotics and the Evolution of Culture*
Melissa Autumn White, "Virus"
Melissa Autumn White, "Viral/Species/Crossing: Border Panics and Zoonotic Vulnerabilities"
T. H. White, *The Book of Beasts: Being a Translation from a Latin Bestiary of the Twelfth Century*
Gweneth Whitteridge, *The Anatomical Lectures of William Harvey*
Joy Williams, *Ill Nature: Rants and Reflections on Humanity and Other Animals*
Rhian Williams, *The Poetry Toolkit, 3rd Edition*
Elizabeth A. Wilson, *Gut Feminism*
Shona Kelly Wray, "Boccaccio and the doctors: medicine and compassion in the face of plague"
Ed Yong, *I Contain Multitudes: The Microbes Within Us, and a Grander View of Life*
Carl Zimmer, *A Planet of Viruses*
Carl Zimmer, "Do Parasites Rule the World?"
Carl Zimmer, "Mammals Made By Viruses"
Hans Zinsser, *Rats, Lice & History: The Biography of a Bacillus*